Alf Baker
WINNING NATURALLY

The Racing Pigeon Publishing Co. Ltd.

Alf Baker — Winning Naturally
First published 1991
ISBN 0 85390 035 3

©The Racing Pigeon

All rights reserved. No part of this publication
may be reproduced, stored in a retrieval
system or transmitted in any form or by any
means, electronic, mechanical, photocopying,
recording or otherwise without the prior
permission of The Racing Pigeon Publ. Co. Ltd.

Text set in 10/11 Century Schoolbook by
RP Typesetters Ltd, for
THE RACING PIGEON PUBLISHING CO. LTD.,
Unit 13, 21 Wren Street,
London WC1X 0HF,
United Kingdom.

Cover picture by Colin Osman

Printed and bound in England
by Hollen Street Press

Alf Baker — WINNING NATURALLY

CONTENTS

I—	THE FOUNDATION STRAIN	11
II—	THE EARLIEST WINS	19
III—	YOUR FIRST BIRDS	25
IV—	UNDER CONTROL, NOT TAME	31
V—	IMPORTANCE OF LOVE MATCHES	41
VI—	BREEDING PROBLEMS	57
VII—	THE ENEMY CAT	65
VIII—	BAD AND GOOD TRAPPERS	71
IX—	WHAT PRICE PEDIGREE?	77
X—	EYESIGN IN BREEDING	87
XI—	YOUNG BIRD RACING	95
XII—	PAIRING YOUNGSTERS FOR RACING	101
XIII—	WAITING AND WATCHING	109
XIV—	IS WIDOWHOOD UNBEATABLE?	115
XV—	BRAIN COMES FIRST	119
XVI—	BREEDING TO TYPE	123
XVII—	OBSERVATIONS AND YEARLINGS	127
XVIII—	AFTER A BAD RACE	131
XIX—	OVERCROWDING DANGERS	137
XX—	MOULTING COVER FEATHERS	141
XXI—	BAD TRAPPING	145
XXII—	EXTRAS IN THE LOFT	149
XXIII—	NO CLOSE INBREEDING	159
XXIV—	WINNING WITH PLUM	165
XXV—	KEEPING THINGS THE SAME	181

INTRODUCTION *BY COLIN OSMAN*

This book is long overdue but never could it have been published at a better time. For a few years the sport of pigeon racing was overwhelmed by short distance Widowhood success and it seemed that money was to dominate the sport. Whatever the publicity given to big money in the national press, there was still a hard core, the majority of fanciers who knew that true success in pigeon racing could not be measured in cash but in the respect a fancier was held by other good fanciers. By these standards Alf Baker, in his prime, was a master, perhaps even one of the greatest fanciers Britain has ever known.

This is a bold claim, even from someone who has known him 40 or more years but the facts speak for themselves. Alf realised that nothing was more boring than a repetition of prizes won and only mentioned them when they were relevant to the story. His years of success were many and covered so long a period that many readers would not have even been born when he won his first Combine race.

What he has to say is just as valid today as it was then. However, to set the record straight I have included a section listing his principal wins and only hinting at the hundreds even thousands of club places he has taken. The wins in Federation and Combine have never been equalled and probably never will be. If he had been flying South Road he would have been proclaimed a National Champion but he flew North Road and only North Road fanciers understand the North Road. South Road fanciers talk of the challenge of the Channel but as Alf points out birds flying 500 miles into London from Scotland probably fly over more water than any South Road bird.

The lack of appreciation of North Road performances also arises from the fact that there are few hot-beds of North Road racing. Apart from London only South Wales has major organisations racing from the North and there are some great fanciers among them too. There is also the North Road Championship Club and the Federations in eastern England that build their programme round the NRCC but they are a special case flying mostly 100 miles shorter than London from the Scottish race points. It is for this reason that wherever possible I have given the birdage, for Alf's Combine wins have been achieved against very substantial competition in the London North Road Combine and even this great organisation has changed so much that young fanciers would not recognise it as it was years ago.

I do not remember when I first met Alf but it was in the 1950s when I was racing from Doughty Street. I was talked into the job of becoming secretary of the New Barnsbury club and still vivid in my mind after all these years is the sight of the old club maps. My job was to prick with a fine needle the position of new lofts and I can remember going to Barnsbury Road to mark a loft. In this short street not far from the famed Caledonian Road leading to King's Cross we already had five members and I was marking the sixth. But looking at the old map, there were pinpricks on nearly every house and probably 30 out of the 40 terrace houses in the street had had a loft at some time or another. Indeed the pub at the end of the street was where the Combine was originally formed. Today I doubt if there is a fancier within five miles of that street. It is all Yuppieland or has been bulldozed for yet more flats. The Barnsbury club has gone, so has the King's Cross and the Chalk Farm and others only survive by amalgamation.

When I first met Alf we were both Combine delegates meeting at a pub in King's Cross that was the Combine headquarters. I need hardly say that the pub has been long closed and there will soon be a Channel Tunnel terminal where it was. I think these examples show drastically how the London North Road Combine has changed since Alf started winning. The radius is still more or less 25 miles from Charing Cross but there are nowadays only a handful of members in the central 10 or even 15 miles. This in turn has led to more changes. Any organisation in England suffers from eastward drift. The prevailing westerly winds favour the easterly members. The western side of the organisation feel they are not getting a fair chance and break away and form their own organisation. In the 1950s if you plotted the centre of the competitors it was not the centre of the radius but North and East of it. When the fanciers were still near the centre of London the eastward drift existed but was limited. Without that centre today the strength of the London North Road Combine is changing and many feel its pre-eminence among south-east organisations is more limited.

In trying to paint the picture of what the London North Road Combine was and is, there is one factor that cannot be ignored, both Alf and I served our apprenticeship under the legendary president Harry Smith. He was the best chairman of any meeting I ever knew and as president of the RPRA rose to the highest offices in pigeon racing. My praise of his ability as chairman is said with the knowledge that he was to fall in disgrace later. At the height of his powers, however, his personality and ability helped to make the London North Road Combine at the time when Alf Baker was at his best a really great organisation. It is also a reminder that Alf, although he was not a pigeon politician, never shirked the work needed to make organisations run. He was always a delegate, a president or a chairman of his local organisations. Would that this could be said of all champion racers!

Over the years Alf and I have held offices in a number of organisations together so that I have had many opportunities over the last 30-40 years to see this side of his dedication to pigeon racing. To name a few: The London North Road Two-Bird Championship Club, London Social Circle, Wood Green HS, Alexandra Palace FC, London NR Fed, North London Fed, London NR Combine. For the latter I was for many years the marking station supervisor and therefore responsible for seeing many of his pigeons off on their winning journeys. When he speaks of the care needed by markers handling pigeons I can vouch for the accuracy of that statement. There was always an eagerness among the markers to handle the legendary Baker pigeons and his anxieties that too much handling or unskillful handling would damage their chances of success.

Over the same 30 or 40 years I also knew Alf on a separate level as editor of The Racing Pigeon and Racing Pigeon Pictorial. In the very early days I was always trying to persuade Alf to take more advertising space. His annual limit was one page in Squills Year Book! What a difference to the breeding studs of today! If I tried to suggest that if he took two pages or even an advert in the weekly, he would have none of it, on the simple grounds that he could sell all his surplus pigeons with ease and bigger adverts would only have meant more cheques to return. By modern standards his lofts were tiny and there was no way they could have been expanded without him losing control and this he would never consider. He may disagree but I think the simple truth of the matter is that he was not interested in the breeding side of pigeons. Of course he cared about how his own birds were paired up but bits of paper simply did not matter.

I visited his loft several times over the years to prepare loft reports and I well remember the first visit when he was keen for me to study his lawn! Perhaps the word lawn is too grand. The photographs do not really give an idea of how small the garden was. It was tiny but he had every reason to be proud of the lawn because it was fine springy grass and as he pointed out it was self-seeding and the shortest self-seeding grass I had ever seen. This was typical of Alf. He had to have grass in the yard for pigeons, so he was going to have the best grass and he was not going out to buy it but cultivate it himself.

It was the same with breeding pigeons. He could have gone anywhere in Britain and probably the Continent and swapped pigeons to introduce into his loft but he did not because, like the grass, the satisfaction was in doing it his way. On another visit I came to the same thought by a different route. I was talking about the number of winners he had bred for other people and said that I could not think of any great performances by others with the Baker pigeons. He looked at me and said that he always told purchasers to follow his methods and he would help them. As he put it to me, they bought the pigeon, but "they don't buy Alf Baker". That is what this book is about, Alf was a master handler of pigeons, perhaps the greatest this century, certainly of his decades. This book tells everything he can about his way of winning. Alf is a wonderful talker and a great thinker about pigeons. My job has been to help get it down in print; it is my trade just as Alf was a glazier.

The first time we worked together was on a series of articles he wrote for the early issues of the Pictorial. They are long since out of print but eagerly sought after because they were a great series of articles. In a newly-edited form they are included in this book. The second source of material was Alf's ill-fated book. He wanted to publish this book himself which was typical of a man who all his life had done it his way. I did offer to help him although he was going to publish it himself. My heart sank and I regretted my offer when I discovered he had already been to a printer and the type was set. I lost my enthusiasm and then other troubles had come crowding in on Alf so that the whole idea died. That material too is incorporated in this volume together with other odds and ends like Squills articles written over the years. There is no way this material could be re-written to make a neat and tidy book, it is just like Alf talking as it should be. Since it includes some material never before published as well as articles written 30 years apart, there are some discrepencies, this is inevitable. In editing the text I have tried to reconcile all this material but at the same time to let the real Alf shine through.

Finally I must conclude on a more sombre note. Alf would not have agreed to this book if he could have avoided it but his recent years

INTRODUCTION

Alf Baker being interviewed by Stewart Guy for a local radio programme.

have been saddened by other events. He always regretted he never had sons who would carry on the loft. It was not to be; he hoped his daughters might follow but it was also not to be. Then his grandson was killed in a tragic accident. Over the same time there had been a long 'campaign' by the taxman over alleged shortfall. The worry of this long drawn action affected his health. The greatest loss was however the death of his lifelong companion, his beloved Kate. She had been his staunch support over the years. They were inseperable and in her quiet way provided a warmth and depth that was central to his life. It is a fitting epitaph for her to say that Alf would not have been the Champion of Champions without her.

Life can also be changed by things that seem irrelevant but are not. Caxton Road was Alf's castle, he knew every blade of grass, every plank in the loft. They started to build the huge Wood Green Shopping Centre literally on his doorstep. The bulldozers and pile drivers were working across the other side of Caxton Road. It was inevitable that he would have to leave. When he finally settled on his own in a new house it could never be the same. Of course he still has a few pigeons, he could never be without them, but it could never be the same.

Colin Osman
1991

WINNING THE NATURAL WAY
Chapter I
The Foundation Strain

How time flies when you become a pigeon fancier. It only seems like yesterday when I built my first loft in 1926 at the age of 12. You long for the racing season to start with the old birds and when that's over you look for the young bird racing. In the winter months you can't wait for the spring to come to pair up and so this goes on year after year. I have lived pigeons all my life and the time has gone all too quickly.

My first loft was built with glass cases, imported from Belgium, at 1s 6d each (7½p) from the glass shop at the top of my road. There at 15 I learned my trade as a leaded light maker and glazier. Inside the loft I put the Belgian strain which I have today — the Stassart. At that time Mons Stassart was the best fancier in Belgium and to me his were the best type of pigeons, not too big, hens especially on the small size, a nice length with good head and eye. What I mean by a good head is an intelligent head. Last but by no means least good feather — the keel well covered and the touch like velvet. I have never seen a good family of pigeons that handle coarsely with dry feather. I know you can help this with a little linseed oil but if it is not there to start with you are wasting your time.

The eye colour of the Stassart family was mostly violet or dark brown. The question of the eye was once put to Mons Stassart. He openly replied he had not the time to study it but he said most of his best pigeons had the purple eye — what we now term the violet eye. My father and a friend, Ginger Taylor — also a good pigeon man — upon my advice wrote to Mons Stassart for six youngsters.

Number One: Red hen, sire 24.2327046 red chequer cock 'Baladin', sire 'Ali', dam 'Andrenople'.

Number Two: Blue cock, sire 27.2309914 'Izard', sire 'Baladin' x 'Florence'.

Number Three: Blue cock 32.2400536 'Job', sire 'Icare', dam 'Gazelle'.

Number Four: Blue chequer hen, sire 31.2405030 'Ibrahim', from 'Debineur' from 'Banknote' x 'Bayadarr' (brother and sister).

Number Five: Red hen, half-sister to 'Icare' from 'Dent-de-Loup'.

Number Six: Blue chequer cock, sire 'Flying Fox' from 'Baladin' and 'Babette'.

As you can see, most of them had the blood of Stassart's best pigeon, 'Baladin'. He won more francs than any other pigeon up to Mons Stassart's death. He was not a good handler and always had ruffled feather at the back of his neck. After he was retired from racing he was missing from the loft over Mons Stassart's restaurant. For five days this caused great alarm and Stassart said he must have gone down the chimney pots. So he had all the bricks taken away from the fireplace and on the fifth one they had taken down there he was, none the worse for experience. This must have been a great relief and shows what a great fancier he must have been. He employed two men: one did all the training, the other looked after the loft, but Stassart selected his birds for the races.

The six youngsters mentioned were never let out and bred winners at eight years old. This proves that it makes no difference when birds are penned up for breeding. I cannot understand why fanciers pay good money for birds only to lose them off the top or killed by wires. I like to see my money in the loft and when you have plenty from them after four or five years then chance them out.

Early Days

Before I left school at 14 I used to do two jobs, one at a dairy scrubbing milk cans for which I got 1s 6d (7½p). Next door was a grocery shop where I used to take orders out on a bike, for this I got 1s (5p), 14lb (6 kilo) of pigeon food 1s 6d (7½p) and 1s (5p) to my mother. I can hear her now trying to get me in out of the rain, she used to say: "You will get your death of cold standing out there looking at those pigeons." She would also say: "I can't see what you see in looking all that time with your nose up against the wire," but all the time I was learning and I am sure this is where I have got my pigeon knowledge, through watching and studying every individual pigeon plus I am sure I was gifted with stock sense. This you must have to be successful with any livestock. At school my main subject was nature study, I never got less than 47 out of 50 and was often called in front of the class to speak on the subject. I used to sit for hours and watch the kingfishers and other wild birds on the river bank near my home and often got a clip on the ear for being away too long.

When I left school I joined the Sunday Morning Club, not affiliated to the NHU as the RPRA was called then, with a half-mile radius. Nearly every house kept pigeons and we used to send 300 birds right

DENIQUE
Winner of six 1st prizes then stock. First London NR Combine Northallerton YBs (5,490 birds). Sent one week after to win 1st London NR Fed, 8th Open London Championship Club Morpeth (256 miles). Grand-daughter of Plum.

through to Perth. No clock, catch and run to where someone was waiting to verify the birds by the ring number which was entered in a book before the race. The birdage fee was 3d (1½p) a bird; most of the short races were done by lorry, for races over 150 miles we used passenger train overnight. I always remember my first thrill, I had the only bird on the day from Perth — a red hen. In 1929, the next year, she was the only bird on the day from the same race point, 9.15 at night, pouring down with rain. She was so wet she looked like a dark chequer. The parents were bought at a Sunday morning auction sale in one of the member's back gardens for 2s 6d (12½p) the pair. I don't think he ever got in the prize-money, under the circumstances he kept his pigeons they were lucky to get home but I knew there was something I liked about them and I thought if I looked after them I could do better. This proved my theory right, I could have sold them back to him for much more than I paid but I am sure he still would not have done any good.

I have often been asked what percentage I place on the pigeon or the fancier. I place more on the fancier — something like 60%; the rest the pigeon, plus a little bit of luck. After all, it's the fancier that makes the pigeon not the pigeon its fancier. The luck you need to miss the obstacles they have to contend with: guns, wires etc, more now with the amount of television aerials there are and these I am sure take a big toll of our birds when racing and the young birds running out.

It's true you hear of a fancier coming into the sport and buying a few young birds from a successful loft and winning straight away. I knew of such a fancier in London who bought six young birds from a good loft and for three years swept the board with them. When he eventually lost them he was not heard of again. He had not got the knack for pairing them up and had little or no stock sense. He asked me down to see what was wrong and the first thing I noticed he was feeding tic beans whilst rearing young birds. I told him I thought this was wrong as I never use tic beans especially for rearing. Also they were almost white, meaning they were only just threshed. I don't suppose I have used ten hundredweight (440 kilo) of tic beans all the years I have kept pigeons but I will leave that for the time being and deal with it when on the feeding processes.

Pigeons and Marriage

My whole life is based on pigeons. I met my wife Kate when I was 14 years old — I was taking some pigeons in a basket to the railway station for a training toss and two girls wanted to see them in the basket, one was most interested and wanted to see my loft. When I had put the birds on the rail they both came home to see the pigeons; my mother made some tea and after about two hours they left. The next evening my sister told me there was a girl at the door wanting to see me, and when I went to the door it was one of my visitors from the day before. I took her into the garden and asked my mother to make some tea and my mother jokingly said: "Are you sure she wants to see the pigeons and not you?"

We soon became very frindly so much so I gave her a pair of pigeons which she kept in a big bird aviary. She used to let the cock out with a note in its ring then he would come to me and I would read the note and replace it with another one. After the cock found his hen was not there I would let him out and he would go back to her. Then one night she came home from work and her brother had got them boiling in a pot! This caused a right rumpus — what I was going to do to him was nobody's business. But it all blew over. I never gave her any more, but we were still good friends and the friendship grew and we were married in 1933 and I still say that was my lucky day when I took my pigeons for a toss. She knew pigeons were my life and she did everything to help me and I am sure she was part of the reason for my success.

We rented an upstairs flat, three rooms and an attic and it meant going downstairs through a kitchen to the garden so pigeons were out of the question. So the birds were kept at my parents. My dad was a heavy drinker and was more times drunk than sober, my mother told me she was always chasing cats off the loft and begged me to take them before the cats had them. So home I went and built a large aviary inside the attic where I lived and took four pairs, six blues, one blue chequer and a red hen — a daughter of 'Baladin'. I kept them in that aviary for nearly a year.

I remember coming back from viewing the house where I lived for most of my racing years and when my wife asked me where the bathroom was I honestly could not tell her. I was not interested about the rooms, all I wanted to see was the garden and that was just what

I wanted for my pigeons. So we took the house and the first thing I did was build a loft. This I think is equal to purchasing good pigeons but I will deal with this separately later on.

Joining a Club

In 1935 I bred two pairs of youngsters from each pair — from these I kept eight, the others I did not like so I killed. The next year I joined the Wood Green Homing Society as a novice. I had won plenty of Sunday morning races, as I told them, but they still insisted I was a novice as I had not won a race under NHU rules. If they were satisfied so was I, and I bred 16 youngsters that year as I told them I would be flying young birds. I used to long for the meeting nights to hear the old fanciers talking pigeons and I must say I was a good listener as in those days some of the best pigeon men in London were members of the Wood Green Homing Society. The 'daddy' of them all was the president, Harry Drayton, known in the bird world as 'Spratty' Drayton. You could take him a chaffinch in a cage and when it sung he could tell you where you had caught it. He had redpolls that he let out for a fly and would return to their aviary.

We soon became good friends as we were both great bird lovers. He was getting on for 60 then but I learned a lot from him and he was a hard man to beat with pigeons. He is the only man I know who had one of his pigeons stuffed in a glass case on the sideboard. He always said she was one of the best pigeons he had known. She won from Banff three years running, being the only bird on the day, and one of two the next year. When he spoke about her his eyes used to water. That's the type of pigeon man he was and why he was hard to beat. When he died it was a sad loss to the sport, especially to the Wood Green Club as he was a great president and leader. I was a greatly honoured man when I was asked to take his place as president.

Getting back to the 16 youngsters these were hopper fed on Tasmanian peas plus a little maize and a titbit of rice, lentils and hemp whenever I went into the loft. I always made sure that the hoppers were filled up at night so they could have all they wanted before I let them out. They used to be gone before I could get outside the loft and go running for three or four hours. When they dropped they would all be inside the loft within a minute where I would give them a small feed of the titbit. A fortnight before racing I took them every morning where I was working, about ten miles. At the weekend

SHY LASS
First Sect, 7th Open London NR Combine Thurso (502 miles). Winner seven 1sts, 1st LNR Fed Thurso.

I put them on the rail to 30 miles. On the job where I was working there were five pigeon men including 'Spratty' Drayton. After six times arriving there with my pigeons four of them said I would sicken them of the basket, but not old 'Spratty' who said you cannot give youngsters too much training before racing and this I agree with, not just to ten miles.

WINNING THE NATURAL WAY
Chapter II
The Earliest Wins

The first race could not come quick enough for me. Off I went with my 16 youngsters on a barrow, gently down the kerb and gently up the other side, you would think I had a new-born baby with me. The first man to look at them was 'Spratty'. He said they looked well and this pleased me very much coming from him. The next day I stood waiting, hoping I would not be too far behind. Over came two birds straight out of the North, one for me I hoped. They dropped like stones when I called them and straight in for their titbit. As I was clocking them another dropped on the loft so I opened up the flap and he dropped straight in so I clocked him. I thought I could be nowhere with three quick ones like that.

When I got up to the club they were asking one another what time they had got. No one asked me but from what I heard I knew I was not far behind. Then old 'Spratty' saw me and asked "What time did you get Young Baker?" When I told them they all looked round and 'Spratty' said, "That's a good one." I said I had two together and someone said, "He's reading his clock wrong." 'Spratty' took my clock to have a look and said, "You — you have got three that will do the lot of us" and so I did; 1st, 2nd & 3rd Wood Green HS. I could not believe it, I never slept a wink that night. I won five more, six out of nine young bird races. After a week or so they were waiting for me to ask my time — this made me very proud so I was never really a novice.

The next year I was top prize-winner and was for years after. Among the 16 youngsters was a blue chequer hen WG337, typical Stassart, a lovely hen she won the LNR Fed four times and ten 1st prizes in three years racing. She would have won three 1sts Fed in a row but after topping the Fed two weeks running, the next week, just as I was expecting my birds, my daughter, aged two, ran outside the garden and I went to fetch her and take her inside. I got in position again and over came a blue cock. As I was clocking him I saw the blue chequer hen 337 sitting tight on her eggs so I picked her up and clocked her. They were 2nd & 3rd Fed beaten by a decimal. It proves you cannot take your eyes off the loft for a second when you have them right.

My good Plum mealy cock was like that. He won seventeen 1sts, five times LNR Fed; he was never more than roof-top high or would come from nowhere and straight in. He was a grandson, bred 1940, off 'Baladin' the red hen I had kept in the attic. The Saturday afternoon when they bombed the London docks — we could see hundreds of planes, there was a dog-fight going on over my house. I stood waiting for my pigeons from Northallerton; my family were in the air raid shelter and my next door neighbour thought I was mad. Over came the Plum mealy cock, 1st Fed!

He won all his races when paired to his mother; as a young bird he showed this keenness for her and just before the third young bird race I let her go in with him for about two hours and he won three 1st prizes including 1st LNR Fed the same way. The next year after pairing him to two hens and transferring his eggs under feeders I paired him back to his mother to race to as she had now gone barren. He won eight 1st prizes in a row in the WGHS. The first race I sent him driving and when I came back from the club I put two eggs under the hen, up he came 1st Club, 1st Fed. After timing him in he went straight on the eggs which I had put there the night previously. On the Wednesday I took the eggs away and he started to take notice of her again and when I came back from the club I repeated the same method; up he came again 1st Club, 2nd Fed. Again after timing him in he went straight on the eggs. I repeated this six more times and that's how he won eight successive 1sts as a yearling.

Army Service

The next year I went into the Army but before doing so I was asked which branch of the service I wanted. I said the Royal Corps of Signals, Pigeon Service, but to my amazement when my papers arrived it was the Royal Fusiliers. After six weeks' training I asked for an interview with the CO, asking for a transfer to the Pigeon Service. After a few weeks I was told I was going to Aldershot on a three-day course on the subject. The following day we went to the lecture room and after listening to the sergeant giving a lot of drivel on racing pigeons I thought I had come to the wrong place. I remember one of his questions was: How would you select a pigeon to fly 400 miles in a head wind? Up went the hands — see that the flights were well up, see that he was firm and hard, see the eyes were bright and wattles white. Then he asked if we'd all done with the questions. I had not spoken before so I said: "Sergeant, if I wanted a pigeon to do what you wanted I would not handle him at all, I would watch his action outside the loft and he would tell me whether he could fly 400 miles." As we left the hall he called me to one side and said quietly, "You will not get it, you know too much," and he was right.

This was typical of the services, if you were a mechanic you were put in the cook-house, if you were a pastry cook you became a mechanic. I remember seeing a sergeant in the Pigeon Service — he worked in our local butcher's shop and the only pigeons he ever saw before he went into the services were dead wood pigeons sold over the counter. I don't say they were all non-pigeon men that were in the service as I know of some good fanciers in our local club that did a good job in that branch of the services. While in hospital after losing my right eye my captain visited me and told me my papers had come through for the Pigeon Service. I hesitated for a moment then he said "or it would mean your ticket". I replied I would have my ticket and look after my own pigeons.

During all this time my wife was keeping them alive on boiled potatoes, baked bread, a little rice and lentils when she could get them. After a few weeks in hospital I was allowed out and to my amazement spotted in the corn shop window maple peas 3d a pound. Inside the shop I went, and asked how much could I have? He said a hundredweight if I required them. As you were on half pay when in hospital a hundredweight was out of the question and I went back to the hospital thinking how could I get enough money to purchase a hundredweight to send home.

That night the usual game of cards was in progress and I joined in. My luck was in and when I had won enough to buy a hundredweight I pulled out. The next morning straight to the shop and purchased the peas. They offered to take it to the railway station for me so I labelled it up and sent it to my home address, also a letter to my wife telling her to expect it. A few days later my wife wrote back to say she had received the peas but they preferred the boiled potatoes, rice and lentils. They must have forgotten what peas looked like.

The day I was discharged the doctor told me after losing an eye I would become very depressed and that I would have to take up a hobby of some kind. When I told him I kept racing pigeons he put his hand on my shoulder and said, "Thank God, I am sure you will be all right,"

THE COMBINE COCK, NU49AX723.
Winner twelve 1st prizes. Three times 1st London NR Fed, 8th Open LNR Combine Berwick (8,000 birds); 6th Open LNR Combine Fraserburgh (5,407 birds). Grandson of Plum.

and sure enough when I got home I had too much to do as the breeding season had just started. The birds looked a picture, neighbours told me my wife used to scrub the loft once a week and clean out once a day. I soon forgot my disability, I still had one good eye and a good healthy lot of pigeons and the main thing, I was home with my wife and three children.

The National Pigeon Service

I then joined the National Pigeon Service which meant you were allowed 15 lb of corn a week plus 1 cwt for breeding eight youngsters for the service. I remember them writing for eight youngsters to be ready by February 1. As I was getting started I could not see what good youngsters would be bred at that time of the year for use on service or for breeding so I ignored their request and made do with the 14 lb of corn plus the diet prepared by my wife while away.

This was 1944: the two local clubs had amalgamated — the Wood Green and Alexandra Palace HS as half the members of each club were in the services. I had eight pairs and they had been off the road for two years so I started training much earlier — a month before racing. Starting them at ten miles and keeping them at that point for two weeks. Then the next week 25 miles; the week before racing 35 miles. I had the Plum mealy cock that was not four years old plus two sons I bred from him the year I went in the services. By this time I had run 'Plum' to three different hens. We had had three races and were a little behind. Then the fourth race I won and was asked if it was with the Plum mealy and replied with his son. Someone said the Plum cock was finished and it set me thinking as he was very fit and well. On the Sunday morning I stood looking at my birds from outside the loft and I saw 'Plum' leave his eggs for his hen to take over and his mother came off the eggs at the same time paired to another cock and they started showing up to each other and were soon parted by the original cock. So on the Wednesday I took the hen 'Plum' was paired to and likewise the cock paired to his mother out of the loft. By Friday he was just following her about as he did as a yearling. The race was from Northallerton, 200 miles, 1st Club, 1st LNR Fed and he went on to six more 1st prizes making a total of seventeen 1sts, five times 1st LNR Fed only when paired to his mother.

I learnt a lot from this; as you can see this was a true love match. The same thing applied with my good hen 'Scottish Lass', she had a flair for her son so I re-paired her to him three weeks before the Fraserburgh Combine, which she won so easily she must have flown it solo. So I could go on with the amount of winning I have done with love-matches.

When I pair my yearlings up I let them go together and pick their own mates, unless I have a pair of yearlings I want to pair together to breed from. I use the others as feeders for my matched pairs and I have never shut two birds up in a nest box to pair up. I let them outside the loft on their own for a couple of hours each morning. When I think they are mated I let them in the racing loft on their own. Then the cock will take her to his nest box then I open the loft front and there is no more trouble. When I visit lofts where birds are shut up in nest boxes I am disgusted. I have seen hens nearly scalped; to me they are poor fanciers just a little time and patience is needed and common sense.

I pair my yearlings up at the end of February, rear one pair of youngsters, then pot eggs and at the end of racing I let them rear another pair of youngsters. This helps to keep them quiet and retards the moult, and also lets them know they are not all pot eggs. I think I must add I never use artificial eggs — I use real eggs boiled as I am sure they know the difference. At least I like to think so — just another one of my fads — and most of my winning is with yearlings in the early races. I like driving yearlings — you often read don't send yearlings driving. I have won most of my 1st Club and 1st Fed prizes this way but there are times when you can lose this way. I never send a yearling driving till he has had five races sitting and if he is driving during the first five races I don't send him. I wait for the next time he is driving and nine times out of ten they win and are hard to beat.

Chapter III
Your First Birds

When you first start to keep pigeons your one thought is to get the best pigeons your pocket will allow you to obtain. First study the different types and make your mind up which type you wish to keep. Study the head, length and size and make sure they are well balanced. I myself do not like big pigeons; I like medium-size cocks, and hens on the small side. These I have found do not need so much exercise to get fit and do not eat so much corn, and fly long races just as well as the big pigeons. In fact, they fly better when the weather is hot.

Having made your mind up on the type of pigeon you wish to keep, write to the fancier for six late breds, four hens and two cocks, not too closely related. Stick to one man's pigeons and do not get them from different fanciers. I am sure you will never get a family of pigeons that way, as they will be of different shape and type and usually these different types and shapes won't hit. Starting with one fancier's pigeons you will have the same type and the following year get some more late breds to put with the others.

Having obtained your youngsters feed them well and watch them grow. I use 60% Tasmanian or New Zealand maple peas and 40% American maize. I like the American maize because it is easy to digest but I will leave my feeding methods to a later chapter. Never let youngsters go hungry. I have always hopper fed my pigeons, not a mixture as I find they pick it about and it gets dirty, just peas and maize. Ignore the rumours that maize can cause canker; I have never heard such a lot of tommy rot. There is no better food for giving energy than maize.

The same time as you purchase your young birds get some ordinary pigeons to make good feeders and take the first round of eggs so you can run the two cocks to each hen for one nest. This way you will find which pairs hit together. Do not take any notice about the fad about youngsters reared by other than their true parents being no good. This is a question I am often asked. For years I had a pair of show birds for feeders and they always reared the eggs from my number one pair which were breeding winners. At one time I had two pairs of African Owls, fancy pigeons, as feeders. I only wish all my racers were as good, you could not get near their nest box when they were sitting and not once could I tempt them off their eggs.

The youngsters you have purchased must be kept for stock and must not be let out. I can assure you it makes no difference for breeding winners when birds are kept in captivity. I had a seven-year-old blue hen that had not been out but bred winners each year. She bred my Dalston Open race winner, also 14th Open London North Road Combine Young Birds. In 1969 she bred my 1st London North Road 2B Championship Club winner, this at seven years old. There is nothing more disappointing than when you start up to find you have a pair that have hit together and bred winners, and then find you cannot reproduce them because you have lost the cock or the hen from that pair. I like to see my money inside the loft.

Using Feeders

In the spring if you find the hens are on the fat side cut the amount of corn down and give them a tablespoonful of Glauber Salts in three pints of water. Dissolve the salts in hot water before adding to the drinker. Make sure you have taken the drinkers away the night before so they will have a good drink in the morning. Repeat a week later, this should get rid of the surplus fat and you will have no trouble in getting your hens to lay in eight days. Pair your feeders up about two days before your stock birds to make sure they have laid a day or so before them. This is most essential as I do not like stock birds' eggs to hatch before the feeders. I would like them to lay exactly at the same time but this is sometimes not possible.

I do not mind putting eggs under feeders that have been sitting two or three days as they will always sit four or five days overdue and will have the soft food when the youngsters hatch out. When you have transferred the eggs under the feeders leave the stock birds together for one day without the eggs. Then take the two hens out of the loft. In three days time put the other two hens with the two cocks. It does not matter which hens go to which cocks; in the first place the breeder will have let you have the right hens for the two cocks.

At the same time put down your feeders to take the next round of eggs from the two new pairs. I always put down a pair of feeders more than I require in case one pair are a long time laying and it helps you to box your eggs around accordingly. When you have transferred your second round of eggs from the two cocks leave them spare for about ten days then bring back the first two hens they were paired to. I always give my hens three weeks break before re-pairing them up, especially in the early months as the egg flow is not as frequent as in the summer. You must try and get for racing two pairs of youngsters from each hen from the two cocks, to find out which pairs hit.

Use the same food mentioned but 75% peas and 25% maize. There is nothing better for breeding youngsters than peas for the main diet. Many years ago I tried rearing youngsters on tic beans; they were the worst lot of youngsters I have ever bred. The droppings around the nest bowl were all wet and sloppy. I had to kill them all and have never again used tic beans for rearing youngsters.

When one reads of well-matured corn this means nothing to me. The only corn which needs maturing is English corn and this can lay about and get contaminated with vermin. I like Tasmanian or New Zealand peas straight from threshing when they are well matured by the climate they are grown under. Do not worry about them being polished; this costs you extra and takes the bloom off, but I am sure it is good for them. If English peas were half the price of Tasmanian peas I would still not buy them as I have had my success with Tasmanian peas and no one will make me change.

The year I tried to rear my youngsters on tic beans I was always filling up the drinkers and I came to the conclusion they were giving too much water to the youngsters, thus making the droppings as mentioned. I found most of the beans for the youngsters in the nest bowl as tic beans when put in water will grow two and a half times the size, and must be hard to feed to the youngsters. I know they have got smaller in recent years but my objection still applies. I do not suppose I have used 10 cwt of tic beans ever since I have kept pigeons and those I made sure were well matured and were at least a year old. But I am not a bean feeder; put maize, peas, and tic beans in a hopper and they will eat them in that order. I like to give my pigeons what they like best but as I have said I will deal with that later.

Designing a Loft

Perhaps I have put the cart before the horse so I would now like to deal with the loft which is of as much importance as purchasing good pigeons. Not so much the tailor-made ones or a new loft; the lofts built by the loft makers look very nice, but not for me. I would have to chop it about and alter it to suit, so it would pay me to build my own to my own design. I have built five lofts since 1926 and with each one, except the last one, I have found the mistakes when finished.

Pigeons must be made happy when inside. My own loft is 16ft long, 6ft wide, 6ft 6in at the front, 6ft at the back in two 8ft compartments one for old and one for young birds with a door at one end. Make sure the door is wide enough to get your basket through. The front of the loft has a 'veranda' running right along the length of it, 3ft high 2ft wide. On the top is sheet asbestos, 4in hanging over the front, this stops the rain from dripping in. The front is in five panels, one is very small mesh wire, the next one plate glass and so consists of three wire and two plate glass panels each about 30in wide.

This veranda is fixed on the front 18in from the top of the loft. In this space I have two long trapping doors 15in high running the length of both sections. These open inwards and through these I let my birds out and return from a race. It is most important never to let your birds fly out through an opening where they can fly straight out, as not only will they hurt themselves by flying into one another but they can hit the sides and damage their wings.

At each end of the loft I have another small veranda, 2ft wide and the width of the loft, 30in high from inside the loft, 2ft high at the front. The top is wire-cast glass, the front and side is very small mesh wire, the back also is quarter wire-cast glass. You may ask why the small mesh wire? This is most important to keep the sparrows out as I am sure they are germ carriers and will peck about the corn in your hoppers and drink from the drinkers. If someone nearby has chickens where they have pecked about in the run they will bring disease to your loft as roup is very common among them. So do your utmost to keep them out.

Never have your drinkers on the floor; the dust from the floor will lay on the top and will put them off form. To me there is no sense in putting down clean water and then in half an hour, with the birds flying about in the loft, it becomes dirty. My drinkers are hung outside where no dust can reach them and are much easier to change. They are boxed in so the sparrows cannot drink from them.

The loft is 2ft off the ground with a concrete base. Make sure your lawn is about 8ft from your loft to stop the rising damp as this is harmful to pigeons. The back of the loft is covered with roofing felt and is tarred every year. Do not put roofing felt on the top, this is inclined to cockle when the sun shines on it and then when it rains it leaves a pool of water. This the birds will readily drink and it does them more harm than good as pigeons love dirty water. There is nothing better than corrugated asbestos on top of wood. Make sure it is 6in hanging over the back for obvious reasons.

Make the inside of the loft as comfortable as possible, as boredom among pigeons is wrong. You will have helped this with the side veranda which they can fly into and sit in front of the glass when the sun shines. Here I have a few box perches, and this is where my drinkers are and grit box. Again these are off the floor for the same reason mentioned.

Chapter IV
Under Control, Not Tame

The first thing fanciers said who visited me was that my birds are very tame but it is not tameness, under control are the right words. The boxes are creosoted each year after the moult and are shut up and the cock during the winter will sit on the perch outside the box. Two weeks before I pair up I put lining paper pinned all round inside the nest boxes in case they are not quite dry so they will not get it on their feathers. After they have reared their first nest of youngsters I take the paper down and they have a clean nest box for the remainder of the year. Creosote is a wonderful thing for keeping down colds. You can notice the difference in the cock bird's wattles when the boxes have been done and all through the summer you can smell the vapour from it. Not only does it keep down colds in the loft but it makes it dark inside; pigeons like dark places to nest in and sometimes pick the most awkward place to make their nests.

I remember I used to have a flat topped corn bin in my old bird loft and each time I came home there was a heap of straw on the top where a pair wanted to nest. For two or three days I kept throwing it outside but they were most persistent and had made up their minds that is where they wanted to nest. As they were two yearlings I thought a lot of I took the corn from inside the bin and let them have it. They were my best two pigeons that year. I can tell you they did look peculiar sitting there but I am sure they would not have nested so well elsewhere.

When I was a boy and had whooping cough or had a bad cold my mother would say, "Stand around the tar pot where they are tarring the road" and it did the trick. My good blue hen 'Scottish Lass', a week before the Fraserburgh race had a dirty wattle, a sign of a cold, so I cut a piece of blotting paper 6in square and soaked it in creosote and put it under the nest bowl. Within four days her wattles were not white but the right colour, pink, to show when a bird is in tiptop condition. She won the race by 17 minutes — she must have flown it solo she was so far in front of the next pigeon.

When writing mentioning boredom among pigeons whilst inside the loft and explaining how I try to keep them happy, as I am interested in all sports, especially horse racing, I was very pleased to read about the then top National Hunt trainer David Barron from Devon, when interviewed by a racing correspondent. To stop his horses from getting bored he left the top door of their boxes open so they can see what is going on in the yard and they can look across fields and watch the sheep. This to me was most pleasing, two men with different sports with the same frame of mind.

Happy in and out of the Loft

Birds must be kept happy inside the loft and out, this is why I have always given my old birds an open loft during the racing season so they can peck about the garden or lay about on the lawn. As I have said, my main diet is 60% Tasmanian maple peas and 40% American maize; this is put in hoppers. Dari, lentils, barley, rice, linseed and a little hemp is thrown on the lawn where they peck about for hours. It is most important that if you do let them go down in the garden you watch what you grow. Avoid rhubarb and hydrangeas; these two are deadly for pigeons. Furthermore do not let them peck any plants; I cover the border of my garden in small mesh net so they cannot get at it.

On one side of the garden I grow a row of spinach and dandelions mixed and on the other a row of spring onions for my own use and I have found they nip the tops off these and they like them as well as I do and they do them no harm. In fact, I am sure it helps to keep down the colds. I remember my old dad when he thought he had a cold coming he would chew a piece of garlic. He reckoned the smell was so great the germs could not get out fast enough!

When they get fed up with walking about the garden, quite close to where I used to live was a cinema and they spent a lot of time flying backwards and forwards. A quarter of a mile away was a gasometer and this is another place they visited. I have seen them on it and they have looked like starlings right at the top. As soon as I stand on the lawn and throw down some small seed they come down. I have no trouble trapping them on race days so I do not mind because all the time they are doing this they are exercising themselves. My old birds do not fly around home. When I let them out in the morning they fly for about ten minutes and are straight down in the garden. I have never flagged my pigeons or made them fly around home.

As I have said, my cocks are of a medium size and my hens on the small side. Also being hopper fed they do not get overfat plus the fact that I race my pigeons hard when the racing starts. They race midweek and Saturday and I am sure if I flagged my pigeons to fly

BLUE CHEQUER COCK, NU68L18727.
Winner of five 1st prizes, 107th Open LNR Combine (7,790 birds) then stock because of his eye and type. His first two youngsters were 1st prize winners.

they would not be so successful. I always think it would give them the impression you are driving them away. I never give my young birds an open loft as I find they are not so easy to control as the old birds at this age. When they have had their fly I get them straight in and always give them a feed of the small seed mentioned. I leave the front open so they can come out again and go down in the garden where they can peck at the greenstuff I grow for the old birds. At the same time I close the hopper so they cannot keep going and having a feed.

I find they are not too keen to go in when you want them to and all the time they are in I have them well under control. For years I have got my old birds and young birds in with a long cane. This you teach them when they are very young. They are put on the top as soon as they get up to the perches, for I am sure not doing this is why a lot of youngsters are lost off the top. They must get used to their surroundings and it gives them confidence when they cannot fly. At the same time I shut the front down so they cannot go back inside the loft. All the time my young birds are out I sit in the garden with them, throwing some small seed on the lawn. When I find they are getting fed up and start to go on the house I get them straight in, never forgetting to open the hopper.

Getting back to the size of my hens, for years one used to read about good roomy hens for breeding. I am sure this is a catch phrase or a gimmick and to be honest I do not know what the words mean or what they are referring to. If it means big hens I am afraid I have no roomy hens; it makes no difference the size of the hens for breeding winners. I could write pages on the very small hens that have been outstanding breeders. One in particular I remember. I bred two youngsters for a fancier, the hen was too small for him he said, mentioning the word "roomy". He was something to do with racing stables and said the best breeders were the big brood mares. The hen he sent back is my 'Blue Girl' 1st LNR Combine Young Bird Northallerton 4,631 birds and is the dam of my good blue cock 'Mick', seventeen 1sts, four times 1st LNR Fed.

Fads and Fancies

At all times I promised myself when writing I would not boast about my pigeons. I only mention this to prove my point. Nothing is more depressing than to keep reading about a fancier's good pigeons. To me he has had all the publicity he deserves after he won the race.

Take the fad about the white streak on the two outside tail feathers, also the dropped outside tail feathers; these were supposed to be no good but these have been some of my best pigeons. Some fanciers do not like soapy beaked pigeons. They say they would not have one in their loft for anything. The best pigeon I ever had was my plum mealy cock, he was soapy beaked almost white and I founded my loft on him and the soapy beaks still come out on the beaks of my blues.

A lot of fanciers hand feed their pigeons and find success. I have always hopper fed my pigeons for several reasons. (1) I do not have to dash home and feed them when I am out. I have seen pigeons arrive at the clubhouse with crops like tennis balls for the race the next day when a fancier has come home from work and hand fed his pigeons. To me these have no chance of winning the next day as all night long they are trying to digest it, thus getting no rest. (2) Hoppers do not get contaminated with excreta and dirt. (3) And by no means least, hopper-fed pigeons have never got a full crop and I am sure when fed this way do not eat as much and therefore do not get fat; a lot of pigeons go to the races overweight to do their best.

At all times I use the best corn money can buy. I know of a fancier in my own club who flies a good pigeon using the cheapest mixture advertised and one begins to think are we not over-rating the corn that we use? One thing I am sure is: the man who masters the art of feeding racing pigeons is on his way to success, and when you have got your own system don't change for someone else's system. A question I am always asked, when it is known that I hopper feed, is: "How do I get them in when I want them?". This I teach them when very young — I never call my pigeons in without giving them something they like; something that is not in the hopper. I have only to go in my loft when my birds are out and they come in and look up to me for the titbit. I always close the hopper about 3 o'clock on the Friday the day before a race and get them in about 5.30, giving them a feed of dari, groats, rice and lentils, all grains which are easily digestible. This gives them a good night's rest in the basket and by morning they have no surplus weight to carry when liberated, and are full of energy.

The Basket, the Pedigree

Some fanciers think that the training basket is the secret of success; to me it is a minor detail. True my young birds get more training than

SCOTTISH LASS NU55ZX9282.
Winner fourteen 1st prizes. Three times 1st London NR Fed, 1st Open London NR Combine Fraserburgh 428 miles (4,032 birds), 6th Open LNRC Berwick. Flying 15 times from Scotland she was three times 1st Berwick and three times 1st Fraserburgh and three times 1st Thurso. Great-grand-daughter of the Plum.

my old birds but this is mainly to teach them to trap. We should think of the old bird training only to get the surplus fat off or to brush away the cobwebs from the brain, not, as most fanciers think, to find their way home. This they have learnt when running out when young, and what you have taught then at this age they never forget. I have heard fanciers when they have been a bit behind at a certain race say they will send them up the road everyday next week, "That will liven them up" but I am sure they have got the wrong end of the stick. They would be better off if they kept them in for a couple of days. This I have found does the trick.

I believe in the survival of the fittest, make the basket your pedigree and you won't go far wrong. "Feed them well, fly them hard" that is my motto. Those that do not come home are not good enough. I have no fads unless the non-eyesign men still think that eyesign is a fad. This I hope to prove to them later that there is definitely an eye or eyes for breeding.

One hears every year of fanciers waiting on the secretary's doorstep for metal rings. Early bred youngsters are something I have never indulged in. We all know the pigeon game is hard work, and when you start breeding this is when it starts; changing nest bowls, scrubbing them out again, cleaning out the nest boxes, and making sure you give them warm water on cold mornings when the youngsters hatch out. If there was any advantage in youngsters hatched the first week in January it would be worth it, and with the seasons changing making the winter drag on to the early spring, it's a matter of luck whether the youngsters are any good. True, I have seen some well reared youngsters when early bred, but that's when we have had some good weather. Furthermore the days are too short for the youngsters to get all the food they need to grow quickly.

Sure enough, this shows keenness in the fancier, but patience is what you want when you keep pigeons. I have always paired my stock birds up the first or second week in February, according to the weather, making sure there is a bit of sun about, I would never think of putting them together on a cold, damp or dull day. I always put down at least six pairs, so I have a nice team when they start to fly and when they go running. I have found with only three to four pairs they hang about and don't put in the work. They seem to wait for the next round before they get on the wing, and when the others start running out they bring them back. They are what I call spoilers.

At the same time as I pair my stock birds up, I put down some good feeders. These are not housed in my racing loft, which is too well ventilated, and therefore too cold. These are in a loft at the back with a glass front, shielded from the cold north-east wind. Note that I mention good feeders; these are of as much importance to breeding winners as good producers. I am sure with a lot of well matched pairings the youngsters have been no good, because the parents were bad or wet feeders, meaning they are pumping up too much water, and not enough good wholesome grain. These youngsters are best put down, as they will never be any good.

My good blue cock, 'Blue Gem', was a wet feeder. This I noticed when he was rearing his first round of youngsters as a yearling. These I quickly did away with and afterwards always transferred his eggs under feeders. He was one of my best producers. There are plenty of good pigeons, but only a few fanciers ever make the top. It does not matter what pedigree your birds have, if you have not got the knack or the good sense to breed good, sound youngsters, all the pedigree in the world will not make them win.

Rearing good Youngsters

I have always said the main thing for success is to be able to breed good healthy youngsters. To me this is one of the joys of pigeon keeping, matching one's skill against nature, in trying to bring out the winning germ, or any characteristics the parents or grandparents have. I cannot wait to see the colours they produce, always looking for that white flight or anything that resembles my best pigeons, knowing that if they have it I've bred another good one. A good fancier knows the quality when they are only 14 days old. I can go back to dozens of good pigeons I have spotted at this age. At all times when I am breeding I stand in my loft listening for that continuous squeaking when the parents are not feeding them. When I hear this I know something is wrong and I inspect every nest bowl to find the trouble. If I see one much smaller than the others I get rid of it straight away, leaving the other one to thrive. I am sure if not noticed in time the other one would not do as well and the parents will do much better with it out of the way.

Groats are a wonderful thing for rearing youngsters, and at all times come to that. I don't think fanciers use these enough. I always put mine through a small sieve to get all the dust from them. Then I put

14 lb in a bin and pour three tablespoons of cod-liver oil on them, mix them up and leave for about three days so the oil has soaked well in. When I have nearly used them up, I do some more ready for when I want them. I give them this last thing at night; the parents readily eat them and they go straight to feed the youngsters. Thus they get an extra feed before dark. Cod-liver oil is a must when rearing youngsters but can be overdone. My birds tell me when to stop it. If I see they are not eating it as readily as they were, I leave it off for a couple of days and start again.

When I wean them off I give them each a halibut oil capsule every other day for three weeks. Those that stop on the floor after this time I kill, no matter how they are bred. I always make a note of the first ones up on to the perches, these are always my best pigeons. I keep a watchful eye on those that take long in finding the corn and water, I treat these with suspicion of feebleness. Never once when I have weaned them off do I hand feed them, they have to fend for themselves or go under.

I wean my youngsters off at 21 days and have sometimes weaned them before this, as soon as I see them eating peas from the pots. I have found they do much better when taken away at the earliest time and helps the parents to reserve energy. Greenstuff is most essential when rearing youngsters. My birds are never without it. In the early part of the breeding season there are plenty of savoy cabbages about. These my birds love and I buy the biggest one in the shop, cut it in half, put one half on the lawn for my race birds and the other I hang up in the stock loft. Plus plenty of fresh oyster shell or limestone grit every day.

Good Cocks not wasted

I only rear one round of youngsters from my race birds. These are paired up at intervals; yearlings on March 1. If I have a pair of yearlings I want to go together I put them outside the loft for three or four hours, then I get them in and put the hen back with the other hens. I do this each morning until I think they are paired up. I have never shut two pigeons in a box to pair them for I have seen hens nearly scalped. To me just a little common sense and patience is needed. The other yearlings I let pick their own mates by shutting the older birds outside and leaving the yearling cocks in. I let the same number of hens in with them for about four hours then I put the hens back. The next morning I do the same thing until I think they have settled down. These I use as feeders for my matched pairs of stock birds.

For years I have only used five or six cocks to breed from and mated them to all my best hens that were not too nearly related. To me good cocks are wasted rearing youngsters and I have put them to six different hens before letting them sit and rear. Ever since I have kept pigeons I have worked around the best cock in the loft. This way I find the hen or hens he hits with and when I do, I pair him to as many hens as possible bred the same way, and also to the dam of the hens. This is line-breeding, the surest and safest way to success.

Many years ago, I was asked about in-breeding or line-breeding and I say as I did then: I have never seen a family of in-bred pigeons that would stand the worst I give mine. I like to keep away from relations as far as I can. For two years I had six pairs of half-brothers and half-sisters paired together and never bred a pigeon worth mentioning. The nearest I go is uncle to niece or vice versa; grandsire to granddaughter, or again vice versa. I am now talking of the racing pigeon for the road and not for showing, for I am sure that pigeons too closely related have a mental defect, weakening the homing instinct, apart from the defect in bone structure, web feet, hook beaks and pipy feathers or two flights coming from the same quill. When someone tries to compare in-bred cattle with the racing pigeon to me this is all wrong, as the racing pigeon relies on its brain for survival on the road.

Chapter V
Importance of Love-Matches

For years I always had a small sack with me when I go to work on the building sites to collect virgin soil from fresh rabbit holes or molehills or wherever they have dug deep holes to put in posts for new fences. Pigeons love virgin clay soil; why, I don't know. It must be the minerals in it. My pigeons have eaten so much that their droppings were nearly all clay!

During the war we had an air raid shelter at the bottom of our garden. This was covered with 14 inches of clay. They ate right down to the metal but the birds flew well. When the youngsters chip out my birds are always looking about the borders of my garden for the occasional little black slug, I often turn over a brick and find them. These they readily eat. I like to keep my pigeons as near to Nature as I can, all this helps to breed good healthy youngsters. In the autumn for years my birds had a feed of acorns. These they eat better than corn, and the droppings are perfect. I only wish acorns would keep till the racing season, as I am sure the iron and oil in them would do the birds good.

As I said in the last chapter the joy of pigeon keeping is the breeding season and then you don't have to count sheep to get to sleep. For years when the breeding season comes around I go to sleep pairing my pigeons up in my head. Always remembering the pairs that bred me winners the year before, these I put straight together again. When there is any shadow of doubt about mating them up in my mind I forget it. When I put them together I like to see them take to one another like a duck taking to water, if they do not after two days I forget that mating and part them.

I have always had my best results from pairs that have gone together as if they were mated to one another the year before. This I am sure is a true love mating and when two pigeons go together this way you have no fear of the hen flirting with another cock. This is something I do not like. I can honestly say my hens are true to their mates and I have seen my hens when paired up and another cock was shown to them, fight him and knock him off the top of the loft. This, as I have said, is an ideal mating, not only will you get the best results on the road but I have found them to be my best breeders, as I am sure they will rear their youngsters with the love they have for one another.

I remember some years ago I paired my number one cock to my number one hen in a small loft on their own. They took a week before they took any notice of one another but eventually they went down on eggs. The youngsters were two horrible little blue hens nothing like the true type they should have produced and I could not stand the sight of them so I killed them. They went down again with more affection for one another and the youngsters were what I expected — true to type. This I have found with late bred youngsters; they resemble the pair you are trying to reproduce more than the first or second round youngsters, proving my point as they are now more adapted to one another. This is something I am always looking for when the breeding starts.

Looking for Likeness

Different characteristics of your best pigeons; white flights, ticks on the eye, anything that the grandparents or great grandparents have, but most of all likeness; this I have found comes out more from the third generation. I have found the hens from a pairing fly better than the cocks or vice versa. My silver blues are not as clever on the road as they are for breeding. I had a pair that were breeding winners each year, but it was always the hens from them. One year they bred a silver blue, the first one they had bred me. As I had stopped racing silvers, the members thought it strange me racing a silver blue. I told them she would win a lot of money in the young bird Combine, this she did and was my best youngster and it was only through knowing the hens from this pair had done so well in the past that she was put on the road.

As I said I get as much pleasure from breeding as I do the racing season. Nothing is more pleasing when you match your skill against Nature when mating your pigeons. They may have bred a winner but I am sure the winning genes come in cycles. If we were to breed one champion every three or four years or a good one each year we would be laughing. I have often gone back to ring numbers and find I have nothing left of a certain year and find also that another year I bred many good pigeons. When one has to rely on six, seven or even eight-year-old pigeons to win, that loft has gone wrong, and they are not breeding the amount of good pigeons each year they should to take their place.

I have never flown a pigeon over five years old and know just when to stop them. One hears every year of fanciers losing their old favourite that has been across the water or to Thurso many times. This can be very distressing and you will lose the best if you keep sending them that once too often. I am sure too many fanciers breed for quantity and not quality; not so much what you have paid for the birds or what pedigrees they have but quality in health. I have seen a lot of youngsters go to the first race that should have been killed in the nest or after they were weaned off. No one is more severe in selecting youngsters after weaning than I and I never give them the benefit of the doubt. Sharp or open vent bones, frets on the flights, tail and cover feathers, wing or leg weakness, these are quickly done away with. When I hear fanciers say it will make up they are wasting their time and money.

Pigeons have always been my second love (sometimes my wife thinks otherwise) but there are two kinds of pigeons I have no time for. The first is one that goes in someone else's loft as a youngster or as an old bird. I have a saying: "The Bakers won't go in" and can honestly say I cannot remember the last one that did. I have had many a pigeon sent back that was picked up flown out, these I treasure as like breeds like. This is why I do not like those that enter someone else's loft when the going gets hard. The second I have no time for is the same pigeon, nine times out of ten. I know the symptoms but not the cure — only the dustbin. These are the words I used a few years ago in the presence of one of the managers of a "Pigeon Curers or Preventatives firm".

I remember reading that a novice had written and asked what he could do with some youngsters that went weak on the legs. They wrote back and told him to buy their leg weakness pills. If he had written to me I would have saved him time and money. No wonder the novice gets confused, there are dozens of pigeon remedies and cures on the market today plus pep pills. If all fanciers were like me these firms would be out of business and there would be no room for them in the sport of pigeon racing, as I am sure there is nothing you can give pigeons to get that bit extra to win. If there was I would have found it!

I am sure that if you start breeding from pigeons that have been cured from this and that, you will never get a family of healthy pigeons and I can honestly say I do not get many sick pigeons. We all have the one-eyed cold, this you sometimes cannot help. Damp or overcrowded lofts can sometimes be the cause or it can be picked up in the racing pannier. This can easily be cured with a few grains of permanganate of potash in warm water, bathe the eye twice a day for three days, putting the bird in a basket away from the others. These I have cured have never done anything worth mentioning after. You will notice the eye you have treated has gone a lighter colour than the other one. Why I do not know, and I do not wish to. As I have said I am not a connoisseur of ailments of pigeons and do not wish to be.

There are so many things one can study on the racing pigeons, the heart, lungs, digestive system, the brain and what have you but it all comes out in the wash when you fly them hard, and all my years as a pigeon man I have spent all my time studying the two main sides of the sport — breeding and how to get the best results from them. Where would I be if I tried to dabble in all of the other things?

Pigeon or Man

There is nothing I like better than to have a few friends round and talk of different aspects of the sport, it is surprising what you can talk about and learn, and always answer every question put to me honestly as I find I cannot talk pigeons without letting some of my secrets out. I remember the late John Kirkpatrick being at a social with several other noted fanciers, we were all asked to say our party piece on our success with pigeons. All I can remember from all of them was you must have good pigeons, placing the pigeon before the man.

Sure you must have the tools to do the job but the tools are no good without the craftsman. So when it came to my turn I had to disagree and based my reply on what I had been saying for years 60% the man, 40% the pigeon. I went on telling them how I prepared 'Red Admiral' to win The Osman Memorial Cup for meritorious performance from Thurso 502 miles. When the meeting was over I was most popular with the other fanciers that were present but not with the other speakers. A few weeks later I was shown a letter that John Kirkpatrick wrote to a friend in London saying what a good night he had had but Alf Baker talked too freely and would beat himself. I have often stopped in my tracks when talking pigeons when the fancier asks me what I put in the water to make them win. I then let him do all the talking as I would be wasting my time, it would be like putting the footings down to build a house and I am past that now. These types are looking for an easy way to success which there is not; only hard work and worry. So the next time you applaud your top prize-winner, it should be not for what he has won, but because

RED ADMIRAL, NU50L4405.
Winner 6th Open London NR Combine Fraserburgh (4,812 birds), 14½ hours on wing, 2nd Open LNRC Thurso 502 miles (3,538 birds), 15½ hours on wing. Winner A H Osman Memorial Trophy. Grandson of Plum.

he has done the hard work and worried, and the good book says you only reap what you sow. How true it is in all aspects of life.

The time I pair my stock birds up is the second week in February and hope for a grand weekend just right for mating pigeons; cold but bright and plenty of sun. The cocks seemed to be waiting for me to put the hens with them. Six pairs had all been paired together the year before and had bred some good pigeons so I just took the hens out of the basket I had taken them round to the stock loft in and put them in the box with their cock. It is always a pleasure to see how happy they are when they first go together again. The hens were just right for mating, no surplus fat. They had been exercised twice a day for the last three weeks, prior to mating, weather permitting, and had been flying over one hour each time. Also they had a tablespoonful of Glauber Salts in the water the week before mating so I was sure they would all lay their first egg by eight days.

Plenty of sawdust and dry sand plus a handful of chopped up straw about nine inches long over the floor. I always used to buy 28 pounds of horse chaff, and, two or three days before they lay, I put a handful in the nest bowl. This helps to keep the youngsters warm when hatched and brings their nesting material just below the top of the nest bowl. I do not like the eggs laid too low down in the nest bowl as I am sure this is the cause of a bad hatch. When the youngsters hatch they will foul the nest bowl and before that the air cannot get round the eggs when the parents are sitting. Just look at the way a wood pigeon builds his nest and you will see what I mean. So I always make sure there is plenty of nesting material in the bowl just prior to laying.

Never put insecticide in the nest bowl with the sand and sawdust — this sometimes can be the cause of infertile eggs. Many years ago I remember reading camphor balls put in the nest bowl and covered with sawdust were good for keeping them free from lice. I tried it but when the youngsters hatched they were trying to get out of the bowl and I found several dead; the fumes had choked them so I never again used it. Some pigeons build up their nests and even after the eggs are laid they still take straw into their boxes. These I have found are my best racers and the ones I look for. Also I take note of the eggshells that are thrown out of the nest boxes when the youngsters arrive. These should be almost in two complete halves and dry, no streaks of blood or other substance left behind in them. This is a perfect hatch and all is well.

Always add plenty of fresh grit each day when you pair up. At this stage they need as much as they can get until the youngsters hatch out. Leave the sitting birds alone, after they have been sitting about a week lift the hen up while she is on the eggs, and see if they are fertile. This we all hope they are but sometimes infertility happens to the best of us for some unknown reason. If you have not got another hen for the cock leave the eggs there for at least another 14 days, do not take them away as soon as you see they are infertile. The egg flow is not as great in the early part of the year as in the summer and you will ruin the hen by making her lay again too quickly.

Pairing the race birds

So much for the stock birds. The yearlings as I told you I pair up the first week in March. The question now is the old birds, those you require for the long races. This all depends on the family of pigeons you keep, as all pigeons do not moult the same. Mine are a quick moulting family so I dare not pair them up before the end of March or they will be too far gone in the wing for the long races. The moult is controlled by the type of season we have; if it is a dry season birds moult more slowly than in a wet season. This is something you will have to work out for yourself. Make a note of when you pair them up and what state the wing is when you get to the long races. I have never done any good in the long races with birds that have moulted their third flight and are on their fourth. These are over the top, for this is when the body moult starts; it is a dangerous condition and you can lose the best.

Rearing youngsters will help to retard the moult, that is why I let the birds I want for the long races rear two first round youngsters and one youngster in the second nest. This also stops the hen from laying too many eggs before the long races. This all helps to conserve energy. Rearing one youngster after the first nest does not take anything out of them: in fact, they put on a bit of weight which you need for the long races as they eat more when rearing. The surplus fat will turn into muscle when required and can easily be disposed of with a good 250-mile fly before the long race.

I never send my pigeons to the long race feeding small youngsters, this is a condition I do not like. Watch your pigeons when sitting. If they look a picture, full of bloom, the colour stands out on the feathers, the eye flashes like lightning and the wing looks as if it is glued to the body, you cannot get them better to do what you ask of them. You will know you have done your best to help them to win. But watch them when the youngsters are chipping out and you will see none of the condition mentioned above. Some pigeons make more soft food than others and this will go sour in the crop while in the race pannier if not disposed of. I have seen birds arrive in this condition and on landing from a race bring it up. When you smell it it smells horrible and it takes a week before they get right again and then I have had to give them milk of magnesia to clear them out.

At all times I can only talk of my own family of pigeons. I can only remember, looking through my record book, one pigeon, 'Red Admiral', which gave me fame and pleasure when feeding a youngster which was 14 days old. We all earmark certain pigeons for certain races, at least I do, and I had made my mind up that Fraserburgh 428 miles was his race point that year. So I sent him sitting about eight days, a condition he had won several times before. They were up at 5.30am light south wind, just what I wanted. I timed him in 6.55pm, 13½ hours on the wing, and he looked a picture when he dropped to take 6th Open LNR Combine, 4,821 birds. My story does not end there, the next morning when I went up to the loft he looked better than when I had sent him. I had four birds for the final race for Thurso a fortnight later that I had kept back but I could not see any one of them beating him if this condition lasted.

As the eggs he was sitting on were now about 12 days old this meant he would have a youngster about five days old. This is the condition, as I mentioned, that I do not like, so I took the eggs away and put them under feeders and put a youngster about four days old under him. Two days later I took the youngster away and put a ten-day-old youngster under him. This had taken all the soft food from him by Monday. Three days before basketing I took the hen he was paired to away where he could not see her. During the day I put the youngster back where it had come from and put it back again with him last thing at night. He wanted to feed it but it had had enough from its original parents. On the Wednesday afternoon, the day of basketing, about 4 o'clock I put the youngster back in the nest box and brought back his hen. He was like a dog with two tails and he never left the box till I took him out to basket him for the race. My money was on him.

Again they were up in a strong north to east wind I knew they would be blown out the west and this meant they would have to face the cold east wind to get back on their line. No one was more confident than me, I knew I would get him and out of the west he came, 14½ hours on the wing, but this time I knew he had been flying. He was just beaten for 1st LNR Combine, 3,588 birds and was awarded an Osman Memorial Cup for Meritorious Performance. It meant more to me than all the Open and Section pools to think one of my tricks had triumphed again and my trouble was not in vain. Incidentally, this was the party piece I spoke about at the London Social Circle where I was supposed to have spoken too freely and would beat myself at my own game. With all the tricks I play with my pigeons it needs the right man to put them into practice. A point worth mentioning: he was lost off the top for a week and I found him lying on the lawn with his keel smashed. The break was half an inch long; after washing it with surgical spirt and smearing Vaseline on it he was running out with the others a week later.

I must point out the hen he was paired to I did not intend flying that year. This is something I always do with the cocks or hens that I hope to do well with. I always pair them to a mate I do not intend to race. Nothing is more disappointing than to lose the mate of your best pigeon during the racing season. What's the good of getting them in the right condition for a Classic race and losing the cock or the hen that he or she is paired to. So mate your best pigeons to pigeons that will always be at home when they arrive from a race.

Form not Luck

I would like as many pennies there are for fanciers that have rubbed up against me for some of my luck. If a fancier relies on luck they will get nowhere! Of course there are some lucky races, pigeons that win that have never won before and never after. We all know the type so I do not wish to bore you with them nor discredit those that win them. It is good for the sport and makes the races go round. I have spoken on some of the luck you need; wires, guns that your best pigeons can meet *en route* especially with a head wind. But the best bit of luck you need is for your pigeons to come into form at the right time.

Nothing in the world can you give them to reach this peak condition but just wait and hope. I have seen my birds a week before a Classic race reach this point and only wish the race was that weekend — they would be unbeatable — and the week of the race they have gone off form; over the top. The same thing can happen after a Classic race, you notice something that was not there when you sent them. The only way we can do this is to get them in the same condition for the Classic races as they were when they previously put up their best performance. Try them always and make a note of this. I have never had to put it down on paper as I have a good memory for the conditions under which my pigeons win — we do not get enough from them to forget!

Never part a pair of pigeons that have flown well together the year before — if you have a cock which you think will breed better with another hen by all means do so; likewise the hen to another cock but make sure they go together again for racing and as I have said fly the best one of the two. Keep the other one at home, I have even stopped racing a good hen or cock when the mate is flying better. I have always found that whatever conditions your birds fly best under, the children will race the same. A good hen that races well sitting eight days will breed daughters that nine times out of ten will do the same. More so the driving cock, I have already said the reason why they go down as yearlings is because they are sent too early in the year. Let them get settled down and get used to it all again before sending them.

Driving Cock System

I have studied the semi-Widowhood system and the Jealousy system or the Two and One as they call it in London, plus the Driving Cock system and find them effective. More so the latter as this does not need the patience or the bit of hard work that goes into the others as the driving cock will come on naturally during the racing season. I now know the ones to put on it, how and when. Take the 'Plum' mealy cock, he was what I call a steady driver. He would not have won all those races if he was a "mad" driver it would have taken too much out of him. The best driving cocks are those that do not chase the hen madly, just follow her about and when she goes to feed he will feed with her and not keep pecking her. These you can send right up to the point of laying, the nearer this point the better.

The mad driver I would not send at this stage. These are the ones you can lose or which go over the top, meaning they have gone racing by with too much excitement and got confused. These I will send only driving three days or just taking notice of the hen. You will notice

DARK DUCHESS, NU68J46422.
Winner of seven 1st prizes including 6th Sect G Avranches NFC and 1st Sect G Nantes NFC. Grand-daughter Sure Return and Shy Lass.

the cock is at the best point of driving a steady driver when his hen is on the point of laying. My cocks are mostly steady drivers all running back to the 'Plum' mealy cock. If you do get a good driving cock you will find his sons will race the same way, I had four sons off the 'Plum' that won over forty 1sts between them driving with hens laying the day they returned from the race.

With most of the systems I fly my pigeons under you do not want a holdover. One day is about the limit, after that you have to keep your fingers crossed. I have won Berwick 300 miles and Fraserburgh 428 miles, well up in the Combine, with driving cocks, two days in the basket. I never send a yearling driving past Berwick but at all times I have half my team that I send sitting about eight days. These I find beat the driving cocks when held over for two or three days. When you do find a cock that wins driving don't be too greedy, remember there is always another year. Twice is enough for one season else he will get cunning too early in life and you will burn him out. The same applies to all your best pigeons that are winning at a certain distance; keep them at the distance they are winning at. Little fish are sweet and I like to win them all. All the time they are learning and if you find they are not doing so well at that distance, send them on. Many a good pigeon has been short lived by sending him to the long races too early in life.

No pigeon can have too much groundwork before the final Classic and if they get a knock in their early life you break their spirit, they lose confidence and sometimes never get it back. You can fly a cock to the bitter end and he can come home flown out several times and he will still breed winners; not so the hens. I am sure something goes inside. After all, she is the one that produces the egg, that is why a good fancier preserves his hens, lightly races them or keeps them entirely for stock. Each year before the YB racing starts I am continuously watching my hens looking for those that I think will keep my family together. These I race very lightly or not at all, as I am sure without good hens you will go downhill.

I know it takes two to breed winners but good hens are the mainstay of a good loft and we all seem to breed more good cocks than hens. This again is in the cycle of pigeon breeding. Years ago I had the best team of cocks that anyone could wish for but not so my hens. There was not one cock among them you could not put on the shelf for stock. But the hens were very scarce for this purpose. Then the cycle changed and I found my hens were better than the cocks, and so it goes. That is why I am always looking for good hens. When I wean my young birds, these promising young hens I say are too valuable for the road. Quite recently I bred a blue chequer hen that had won two good races from the distance. When I saw what she was breeding I stopped racing her knowing she would have won again. One of her sons won as much as I would have wanted her to. These are the things you must look for to keep at the top. Getting there is not easy but staying there is much harder.

The Feeding Question

In one of the earlier chapters I said I would leave the feeding processes to a later date. This is the most common question I have been asked ever since I have kept pigeons. I have already said some fanciers have found success by hand feeding; some, like myself, by hopper feeding, leaving the corn there all the time so they can eat it when required. But I am sure hand feeding does not mean throwing it about all over the floor. This should be fed into hoppers in trays until they have had enough to eat. Then take the trays out of the loft. This is where a mixture can be used.

If you leave it there all the time, as I have said, they would pick it about and sort out the best or what they like best first. Some use a dear mixture and some use a cheap and still win but at all times it must be good sound corn and not damp or musty smelling. This is something you do not have to worry about when you use New Zealand or Tasmanian maple peas as they are well dried by the climate that they are grown in.

I have never been a wheat feeder, not since I have not been able to get the Canadian brown Manitoba wheat, hard as a rock. English wheat is inclined to make them loose, Australian wheat is the next best to the brown Manitoba. Wheat is the first thing rats and mice go for in the barn and as I have had my success without it I leave it alone for that reason. Mouse and rat droppings are something I am always looking for when I buy small seed, linseed, dari, lentils, rice or hemp. If I see any I throw the lot away; that is why I only buy this in small quantities.

You will notice I did not mention canary seed. I know quite a lot of fanciers use this but I do not like it because there is too much wood, meaning husk, on it. Even the canaries take this off before they eat it. I know pigeons need roughage, that is why I use a little malting barley. Again I make sure the rats and mice have not been at it when I buy it. A few tares used as a titbit will all help to vary the diet but I have found these of poor quality lately, not like the Scottish ones before the war. Groats are an excellent food for breeding and racing, being easy to digest, but make sure you sieve them to get rid of the dust. A couple of handfuls of oats thrown over the lawn will help to keep them quite happy. Oats are groats without the husks on and again this is where they will get the roughage required.

Cornflakes crushed up and put in gallipots; Scots Porage oats the same; breadcrumbs, these you can buy at any baker's shop, you only need three or four pounds, all these they like. We never throw any bread away. This is put in the oven till lightly baked just a light brown then crushed up and again given in pots. I often buy a loaf of bread, let it get hard, hang it up inside the loft and they will peck all the inside out of it. Wait till the loaf has got a bit stale. I have often had it hanging up when expecting visitors and have taken it down else the word would get round that: "Baker feeds his pigeons on bread!". They would say I raced on the corn tin, to the novice meaning keeping them hungry. Not me but I know quite a lot of fanciers that do but when they get to 200 miles they do not time in; before then when the wind is on their nose. These are poor fanciers and I would like to give them the same treatment.

As I have said, different fanciers have different feeding methods and find success. I remember a few years ago a very good South London fancier who kept pigs. He used to mix the pig meal up very stiff and no sooner had he filled up the trough for his pigs than his birds were down eating it and he won more than his share of races even from Lerwick. I never saw a fitter or more active team of pigeons. I remember seeing him put a handful of something with the meal and I asked him what it was and he said dried seaweed. Being curious I asked what it was for. He asked me if I could see any running noses among his pigs and I honestly could not. "That's it," he said, "it keeps them free from colds." I took some home and put some in my mineral salts plus calcium phosphate for strong bones and have used it ever since. Thanks to him I have a healthy team of pigeons. He has long since been gone and I have always admired him for good stock sense.

The saying is: "There is no substitute for wool". This phrase should be used about pellets: "There is no substitute for good sound corn".

Whatever some fanciers wanted to try them for I will never know. No one knows what type of corn goes in the making of them. They may be all right for chickens but that's not what we keep. I would get better results from the diet my wife prepared while I was in the Services. At least I would know what I was giving them.

If I live to be a hundred I would still be learning about the racing pigeon, but since 1926 I have studied every way to get that little extra to win with them which I hope to pass on and make someone as proud as I am with my pigeons.

Chapter VI
Breeding Problems

I once had a good stock hen brought round to me, she was picked up from the railway bank near my home by a boy who lived next door. I knew straight away what was wrong with her, wing stiffness, or leg weakness and the day before she had just laid her second egg. This sometimes happens to hens after laying. I remembered the previous year after this hen laid she could only just fly down to the garden, but after two days she was all right again.

I have had some hens after laying, that could not stand on their feet for a couple of days, but they soon pulled round, and are none the worse after. I am sure this is inherited, as the dam and granddam of this hen were the same, but they were good racers and breeders. This is only caused, I think, through the egg passing through the two vent bones and touching the nerve, which causes a form of paralysis in the wing or legs, so not to worry.

I am sure there is more in colour mating than fanciers care to think. Blues and blue chequers paired together are practically the only two I would mate of the same colour. I do not pair two mealies together, these I put with blues or blue chequers, or dark reds. I do not like two red pigeons together unless one is much darker than the other. They should go with the blues or blue chequers to get the depth of the colour. I do not like light mealies; my best mealies have been strawberry mealies. These you will get if you pair reds and mealies as above. True, a good pigeon cannot be a bad colour and I have had some good mealy cocks as producers, but not so the mealy hens. This is why the breeding side of the sport is so interesting.

These are the things I bear in mind when taking in a cross. I know when you have a family of good winning pigeons, one is always afraid to bring in a cross. This was my case after the war; my pigeons were winning out of their turn, and I could see I would be getting too close when mating. Not having taken in a cross before I was worried, for I had read that when taking a cross you weaken the blood of your own family. True, it becomes fifty-fifty, but since then I have found different, and now have no more worries when thinking about a cross. I had only four crosses since 1948, and each one from the first cross has been a champion. I always bring in a cock. Why cocks, you may ask? For my own three reasons:

(1) A good fancier preserves his hens; (2) You can run the cock to three or four hens in one season, thus finding out how good he is in a short time; (3) I have a knack of picking out cocks rather than hens, for the intelligent look shows out more in cocks than hens.

The four crosses have all been Belgian strains: Ameels, Bricoux, Sion and my latest, the Delbars. You may ask why go to Belgium for my cross? As you may know, the base of my family is the Stassarts, and as Mons Stassart was one of the best fanciers before the war, so some of the blood could be in the strains mentioned. But the main thing is they all looked like my own pigeons, even had the same eye colour, and were all blues, the same as my own. When I put them in my loft, apart from the ring, you could not tell the difference. This to me is the main object when taking in a cross, I make sure they are the same colour as my own family, the same type and size, therefore resemble my own family, and last but no means least they have the same colour eye. At all times they have gone to my best hens to prove their worth and as I have said, the best pigeons have come from the first mating of the cross.

I am sure a lot of good hens go to the wrong type of cock in respect of size, type, length and balance. Although I have no big pigeons I would not put a big cock to a small hen or vice versa. When I first started keeping pigeons I had one type in mind, the type I liked, and, as my family of birds all came from the same mould, I produced that type. True, I do breed an awkward one now and again, or a henny looking cock.

Disliking a Bird

One year I bred a chequer white flight: "What a lovely hen" I kept saying to myself, but early in the spring of the next year it turned out to be a cock, I changed to disliking it like the awkward ones, but when racing started I could not keep it out of the clock and it was a winner of many 1st prizes including 1st Fed. But I still did not like it! I find when you take a dislike to pigeons, no matter what they win, it does not alter your mind about them and for all the winning the awkward and my odd-looking cocks do, I would never breed from them. Once you do your loft will be full of this type, and I like to win with good-looking pigeons, or put it this way, the perfect ones.

I would sooner be 2nd Combine with one I liked than win with one

I didn't. This alone gives me great pleasure. I know some fanciers do not mind what they win with, as long as they win, and some cannot tell the difference between the nice-looking ones and the mean-looking ones. I remember visiting the loft of a fancier who was doing very well, but when I saw his pigeons I did not like them; mean-looking cocks and hens, small wattles, small round heads, short beaks. They were flying well, but honestly I could not keep them; they were not pleasing to look at, and surely we must get a lot of pleasure by looking at our pigeons. This fancier had not got in his mind's eye a type of pigeon when starting.

I have seen brood mares and their foals at the Newmarket stud. I have seen some of the best greyhounds in the country, but nothing is more pleasing than to look at half a dozen blue cocks that have just completed the moult. To me it is like looking at bags of gold dust. Early in April I rung the 12 youngsters from my six pairs of stock birds put down on February 14, and they almost always looked very nice, no small ones among them. All the years I have kept pigeons, one job I hate is putting the ring on youngsters. Why, I do not know, perhaps I am afraid of hurting them and so I sometimes leave it to the last minute and have to put Vaseline on the leg.

As I have said many times, mine are not all big pigeons, but all have good strong thick legs. I know we do not expect them to walk home but this is something I like. To me it shows good strong bone structure, so don't leave youngsters too long before you ring them, I am sure if you hurt them while doing this, they do not forget, and sometimes this will be the cause of bad trappers. The best time to ring youngsters is in the morning, when the parents have not fed them. They are much easier to handle, and you don't make them sick for when the crop is full you are likely to squeeze it and make them bring back their food.

Early training; early races

When you see a batch of trainers coming over the top of the loft, wind cold north-east a month before racing, this is like breeding very early youngsters; that fancier is taking a chance. I would never think of training pigeons that time of year with the wind in that direction. Pigeons do not like a cold east wind, and many a pigeon has been ruined through training on these days. They have not had the amount of sun to get them fit and if they get a bad toss at that time of year they could be finished for the season. I am sure these are from the fanciers who have been hibernating since racing finished, and are now becoming keen. True, it is the summer when the race results come out but races are won during the winter months. Pigeons can almost look after themselves in the summer; it's in the winter when they need more attention. But these fliers seem to think they can catch up by training early.

The same applies to the Open races that are organised early in April, the majority of fanciers that support them do not even consider the weather previous to sending or on the weekend of the race. I have heard them say after the race that they have lost Thurso pigeons that have flown well the year previously. I would never send a pigeon of this calibre to an early event. I may send a yearling or two, but as I have said they would have had to have the sun on their backs previously, and it would have to be warm that weekend. I have seen my wife undo every knot of a parcel; not me, I cut the string and rip it open. She would say you have got no patience. How right she was! But one thing I have got patience for and that is pigeons, and that is all I need.

Pigeons must have some animal protein, that is why you will see them eating worm casts off the lawn. Years ago I used to buy a pound of beef suet and chop it up and give it to them in pots and they looked forward to having it. I now use cod-liver oil to take its place. This can do more harm than good, not so much by overdoing it and giving them too much, as I have said, they will not eat it when their body has had enough and does not require any more. It is then I stop giving it to them for a couple of days.

The harm does not come from this but from the way it is given. I would never put it on my corn even though I hopper feed. If you do, when the birds fly about in the loft the dust and even the small down feathers, from the loft will stick to the corn, as peas, beans, maize and barley etc will not absorb the oil. Some fanciers put it on their small seed for trapping. I can just imagine a tin of small seed with cod-liver oil put with it, say a teaspoonful to six pounds, it would all stick together, as none of the small seeds I have mentioned will absorb the oil. When the birds start to eat it, it will get all over them, and they will look terrible, I remember seeing a fancier who had even put it with his grit, I can see them now, what a state they were in!

In my opinion the only way to use it is on groats; these, in two days, will soak up the oil and therefore it will not pick up dust or get in

GAME ONE. NU68L18724.
Winner of seven 1st prizes, 1st Stonehaven, 24th Open London NR Combine (6,274 birds); 1st Fraserburgh, 22nd Open London NR Combine (4,438 birds). Son of Scottish Lass and Margurite.

the feathers and, most important, it will readily be eaten, and will not lay about and go sour. For years I used to give my birds the water from smoked haddock when boiled, but I do not think many of the haddocks are smoked, nowadays they soak them in a yellow brine. They cook the same but the water from them is different. All this contains animal protein which I have said pigeons must have. My wife never threw away the water from boiled greens; I take the drinkers away at night and fill them up midday with the water from the greens. This is a wonderful tonic for them, and will fetch out the plumage. Best of all are dandelion leaves plus the root. These are washed and chopped up and then boiled. The water from them is added to the drinkers. Never give a pigeon cauliflower leaves; these scour them out and are too strong for them.

Slow layers

One year I made a mistake with my best yearling cock, for I paired him to a slow-laying hen. He had been paired to the hen for over ten days, she had not laid and he was beginning to look fed up. She laid in eight days the previous year, but had a hard race from Thurso, proving my point about taking hens producing your best from the road. If she had not laid in the next two days I would have tried putting a warm egg under her in the evening. This is something I do not like — pairing my best pigeons to slow layers — it takes too much out of them and you cannot be sure of getting the cocks in the right condition for the races you want them for. When hens do not lay at the right time, seven or eight days, your calculations can go all wrong.

This is the only reason I do not like slow-laying hens. It has made no difference in their performance on the road. My good blue hen, 'Scottish Lass', one of the best hens to fly into London, always took over ten days to lay, not only her first pair of eggs but all other pairs after. Her youngsters were weaned off before she laid her next pair of eggs. All other hens laid when the youngsters were 16 days old. She was never a keen racer when she had youngsters, but get her sitting eight days she was always hard to beat. I sometimes used to think it was beneficial to her as, by the time racing was over, the others had laid two pairs more than she. This could have helped her to reserve energy for the long races, at which she was best.

What a relief when you have got the birds all nicely paired up and settled in their boxes. Every year this to me is one of my most worrying

Breeding Problems

times. I do not get much fighting over boxes; my cocks have had their nest box all through the winter. As I have said, I have no nest fronts. This all helps to make it much easier. I know nest boxes with nest fronts look very nice, but I am sure this causes fighting and smashed eggs. Not so without the fronts; the cock can easily be knocked out even when he just flies on the front of the nest box. The hen sitting will let him know he is not welcome even before he enters the box. Not so with nest fronts; once he goes in it is not so easy to get out. That is where your eggs will get smashed. True, I still have nest fronts, but learned not to make this mistake many years ago.

I do not like pigeons fighting when it can be avoided, as it is surprising the amount of damage and harm they can do to one another. I have seen two cocks fight one another to a standstill, and it has taken them weeks to get right, if they ever do for that year. I have found that when two cocks fight early in the year they hate one another for the rest of the season, and the trouble can flare up again any time during the season. I usually find I get this sort of trouble from birds I pair myself. I am afraid there is a lot we do not see during the winter months when the birds are parted, certain cocks have been "winking" at certain hens through their wires and then we pair them up differently. This, I am sure, is the trouble.

I never get any trouble from the pairs I have let select their own mates. A lot of fanciers call the fighters fools saying they should know they have gone into the wrong box. I have found it different when the time comes to prove it. One is always talking of cocks in this respect, but two hens can be just as troublesome, as I have found in the past. I stand in my loft for hours when I first pair my race birds up, and when a cock or hen flies on the wrong box I push it off. It is no good taking it out, it will fly straight back; just a quick flick with the hand several times and it will know it is doing wrong. I find the trouble comes mostly from yearling cocks, not so much because they are fighters or troublemakers but because they are bursting over with joy and energy after the hen has laid. Now they have their own nest box for the first time, and to me they are showing off.

Be patient with them and never lose your temper; these to me are your best pigeons so do not treat them as fools. As I have said they will prove this when the time comes and your patience will be rewarded. I now have to wear a mask when I clean out because of the pigeon dust. It is surprising the amount of dust in the loft when you clean out. After over 60 years, the amount of dust I must have inhaled over the years has begun to tell. My father and grandfather suffered with bronchitis and this has been handed down. I would advise anyone, young or old, who keeps pigeons and suffers with bronchitis to do the same; the pigeons soon get used to it. Nowadays I keep a small team.

One strange thing whenever I have a good year is that my cocks are treading one another. To me this is a good sign of fitness, especially just after mating. A friend of mine used to say he did not like the ones that got down for the other cock to tread, but I have made note of these and they have flown just as well as the others. I remember a nice light chequer cock he had, and I saw three cocks tread him in turn. My friend also saw it and I remember him saying the light chequer will be no good, so I asked him to catch him for me. I wrote his ring number up in his loft, and I said to him you will not only see his ring number up there, but also on the race sheets at the top. Sure enough it was, not only on his club race sheets, but twice in Combine taking a prominent position each time.

Chapter VII
The Enemy Cat

I have already spoken of the many enemies the racing pigeon has, the trigger happy man with a shotgun, wires etc. But the real enemy is at home, the dreaded cat. On both sides of my garden, the fence had three feet of interwoven fencing above it, with two feet of mesh wire on top of that. The top of my loft had three feet of wire mesh all round. Some years ago I put four climbing roses on the other side, this stops the cats more than the wire does. You can buy roses that have plenty of thorns and cats certainly do not like them.

Every year when I gave my birds an open loft I went round and checked all the palings on the fence, and repaired the wire. Ever since I have kept pigeons I have done my utmost to keep the cats at bay. They can ruin a round of eggs if they get in the loft as by the time you get the birds over the scare, the eggs are chilled.

If you have any out-house near your loft where cats can sit at night, and peer into the loft, making it inaccessible to cats or the birds will not get the proper rest they need and if you have a cat run, meaning where they go back and forth, put the rose bushes there, this will stop them.

Seasons can vary, one year I had only one clear egg, but can tell you I had had my fingers crossed because of the cold north-west wind we had. There is an old saying "Wherever the wind is in the last week in March, that is the prevailing wind for the year". So I predicted there would be some fast erratic races on the North Road. Whenever I pair my pigeons up and there is a cold north-east wind I keep my fingers crossed.

Three years earlier we had the same weather when I paired up my birds, and I had seven clear eggs. I put it down to this — it is obvious the hen will not get down for the cock as much as she should. I am talking of young pigeons not the very old. Infertile eggs can sometimes be caused by old pigeons making too much feather around the vent and not making contact with one another. These I inspect before pairing, and trim with a pair of scissors. I do not rely on old pigeons to breed my racing team, at least not two old pigeons together. It is true that one hears of a very old pigeon breeding a good one, the sire of my 'Blue Girl' was 11 years old, but the hen was only two, so as I have always said there is no book of rules to pigeon breeding or racing.

Finding the Good Ones

There are several ways of telling healthy youngsters when weaning. First, I always inspect the state of the nest bowl in which each pair has been reared. These should be almost the same as before the youngsters were bred, just a few droppings around the outside of the bowl and the scalings from the flight feathers when breaking through at the bottom, also the bowl must be very dry. There is not too much wrong with the youngsters that leave the bowl in these conditions.

I always inspect the vent, this should be well covered, without any yellow droppings stuck to the feathers around it. If dirty I mark it down as doubtful, and I always keep a watchful eye on them. I do not like youngsters' tails to stick up when you handle them, like a cock robin's. This is the only way you can tell a weak-backed pigeon, the tail should point down slightly when in the hand.

I know some youngsters, like old birds, do not like being handled and will keep bringing up their wings, therefore one cannot get the feel of them. I never let my hands be the guide to a good pigeon. Many a good pigeon has proved its master wrong. For example take the 'Plum' mealy cock, I was always trying to get weight on him, by hand feeding him with maize. Every time I put him into a race I wished he had more weight like my good blue cock 'Ragamuffin'. He got his name by the way he handled, and by the hand he was never a 500-mile pigeon, but he proved me, and several others, wrong. Liberated at 5.15am and clocked in at 9.27pm on the same day from Fraserburgh, 428 miles, 5,407 birds competing, with only 18 home on the day. Two weeks later he was 14th Open LNRC from Thurso, 502 miles after nine days in basket. He was the lightest cock in the loft and never weighed over 14½oz.

I used to weigh my birds on a pair of scales before and after the race. The light frame pigeons hardly lost any weight, but the heavier ones did. I remember marking a blue hen for Thurso, the only one a fancier had sent, it was not much bigger than a blackbird and was pooled right through. When I got it out of the basket to ring it I said to myself if the race was hard the bird would not make it home. It was hard all right, 15½ hours on the wing with only two birds home on the day, and this bird was one of them! This certainly made me think and it taught me a lot.

The Enemy Cat

No one knows a 500-mile pigeon until it has been tried at that distance. All the best 500-milers I have had would never win in a show, the judges would say that they were too open across the back, meaning the cover feathers are wide apart. These, I found, have always been my best distance pigeons.

I like to think that with the cover feathers wide apart it is much easier for a pigeon to lift his wing up, as the air can get through from the secondaries and not the body. This will not alter the downward thrust, as the secondaries will close towards the body thus lose no air space. I am sure that the upward movement of the wing, when in flight, is harder work than the downward thrust. This can be seen by the slow-motion film that *The Racing Pigeon* had years ago. Though as I have already said, the only way to find a 500-miler is to send it.

Do not ask a boy to do a man's job as some fanciers do. I know some yearlings have put up some good performances from 500 miles, these are the ones that we hear of, but we do not hear of the ones that go down.

Narrow headed pigeons are another of my dislikes. I like to put my thumb just over the top of the wattle and still see the eye-cere. I remember being visited by one of the crack Belgian fanciers. He picked out nearly all my best pigeons by the eye-cere, choosing those with little or none at the back. He also used a sliding rule system, by putting the finger next to the thumb under the wing butt and using the thumb as a stop to find the width. Nine times out of ten they have been right. Unlike him I leave the basket to do my sorting out, it is surprising how many followers you have in the loft when you start training. I have sent my pigeons out several times to 35 miles and they have all dropped together. But take them half that distance and put them up in twos or threes and see how many you have when you get home.

The strangest part is those that make straight for home, as you think, when you let them go, are not there, and those that seem to go the opposite way are home when you get back. A lot of pigeons rely too much on others and are what I call followers. It takes a good pigeon to fly a distance on its own when others have finished their journey, and they don't get the credit they deserve.

One hears a lot about overfly and wind. I know the wind plays a big part in any race, but the drag is a far greater factor. I would sooner fly short but on the main line of the bulk of pigeons than get overfly away from the drag. This has been proved many times in the Classic races, and I have never bred a bird that could beat the drag. In London they call it the Golden Mile, even the property prices rise in this area when a pigeon fancier is house hunting. Before moving quite a lot of pigeon men study all that I have mentioned, and within a few years make a name for themselves, before the move they could be unknown, again proving my point regarding the drag.

Youngsters on the loft

Youngsters must be let out at every opportunity, nothing is more depressing than to lose a nice youngster off the loft. I never lose any sleep when I lose youngsters while running or racing, but I hate to lose them off the loft. I can honestly say I do not lose many, unless a crow flies by and frightens them, but if they have been out when very young they will work back. I let them out each morning at about 6.30, and get them in before I go to work. Then I let the older birds have an open loft, I never let young birds and old birds out together. My old birds roam all over the place, this would give the youngsters bad habits, and as most young bird races are won on the trap, I like to keep them well under control. When the old birds are flown out with the youngsters they are inclined to get on the wing too early.

I like to hold my youngsters back till they have cast their first flight and the eye has changed, all the time they are getting a good look around outside. Sometimes they get on the wing before the keel has hardened. When youngsters are lost off the loft before the eye has changed there is very little chance of getting them back. For my novice friends I would like to explain what I mean by the "changing of the eye". When youngsters are weaned their eyes are a plain, dull, brown and you cannot see the iris. After five or six weeks you will start to see the change, the pupil and the iris colour will start to show. By eight to nine weeks the eye will have completed its change. Youngsters at this stage are safe, most that I have lost off the loft have been between the eye change, as they are nearly always the best for they are the most advanced ones.

I know all of us cannot wait to get them on the wing but if it is done too early this can ruin them, I have had them come back after several days merely skin and bone. It shows guts and courage, but they need nursing with plenty of easily digestible small grain such as groats. But most of them have been no good after, something goes wrong inside and they go down the first time the going gets hard.

Youngsters from some pairs simply cannot be held. I remember my good blue chequer hen 337. For two years I never held a youngster from her, no sooner had they flown out on to the top of the loft than they shot off as though something had scared them. They would fly higher and higher and I never saw them again. So the next year I put them on the top of the loft for a week, six days after weaning. When I let the other youngsters out I kept these in until the eye had changed and they had cast their first flight. The first time out they did the same thing as the other youngsters from this pair, but I got them back late that evening and they were no more trouble. Every bird this pair bred me proved to be a winner, but I had to hold back until their eyes changed, and this proved my point — it's the most advanced ones that you lose off the top.

We all start the season with a nice team of yearlings expecting big things from them, but as soon as their youngsters reach about 14 days old you see how wrong you can be. Those that show signs of distress while feeding, for example.

I had two nice chequer hens that I thought a lot of, but now their youngsters are four days old you would think they had flown 500 miles. I am willing to bet they will go down when the wind is on their noses. This is something I take particular notice of, pigeons that can't stand the strain of feeding youngsters have not got the constitution to stay on the wing to fly 500 miles, and will never make champions.

Soft Food Danger

I have written many times on the best conditions to send pigeons to a race, but watch out for the pigeons who have turned in their eggs two or three days prior to basketing for the next race. This is one of the most dangerous conditions to race them no matter how good they have flown previously, they will do no good at this stage. Some fanciers think pigeons only make soft food when the eggs start to chip out, pigeons make the same amount of soft food even on pot eggs; you cannot get away from Nature. As I have said before this must be diposed of or it will go sour in the crop, this is partly the reason why they should not be sent during the soft food period. Also they are most disappointed, the eggs have not hatched and they become disheartened.

I know that one likes to get in as much ground work as possible before the Classic races. If I see a pair that are about to leave their eggs three or four days prior to a race and I want to send them, I give them a three-day-old youngster. This will help to take away most of the soft food and it will sometimes come up trumps, but one thing I do know, they will home on the day. While they are away I clean the nest box out and replace with a clean nest bowl for them to go down again. Those I do not send that have turned their eggs in I give them a half-teaspoonful of Milk of Magnesia from a nose dropper, to save them getting a sour crop. Having made many mistakes in my early days I now give the pigeon the benefit of the doubt.

In my early days I used to note on paper the birds that I thought were not to my liking or quite right when basketing for a race, and nine times out of ten they were late or next day arrivals so now, if in any doubt I don't send them. The birds I am referring to are those with feathers on top of the wattles raised, loose in feather, dirty or greasy wattles, those that are listless while outside the loft. All these birds have no chance of beating a fit pigeon.

I am sure one can learn more by watching the actions outside the loft than in any other way. Cocks that cannot keep still and are looking for trouble, those that keep taking off including hens that take off and fly on their own for 15 minutes, hens who drop their flights below the tail and keep flicking them back. There is nothing wrong with such pigeons.

I only pool three or four of my team, those that I prepared for the race. Many a time I have gone to basket my pigeons up for a Combine race and have spotted a cock or hen in the condition I have already mentioned. I have pooled it right through and this has always paid dividends.

The right time to watch cocks is between five and six in the evening when they have changed over from their eggs, or an hour later in my case when the sun cast a shadow half-way over my lawn. I walked my pigeons into the shadow with the green background, and every bit of colour in the plumage stands out.

I would not think of sending a hen to a race, even if they go away for a race the next day, when I see the cock start to drive her. I have seen hens when they have been race rung go straight to a corner of the basket and "lay". Although you may think she may not lay for a couple of days there is always a chance of a holdover, and I am sure a good fancier would never think of sending a pigeon in this condition.

Chapter VIII
Bad and Good Trappers

Every year as soon as the racing starts you will hear fanciers moaning about the bad traps they have had. They blame their wives, or someone else for giving their birds to much food before basketing. All I hope is they do not say to those who try to help them, what I hear them say at the clubhouse. If they do, they will soon wish they had not, as a little help in the pigeon game goes a long way, and I would have sooner my wife had overfed them than not at all. So take my advice and keep quiet, for pigeon fanciers' wives have a lot to put up with.

Apart from the many hours spent at meetings and with the birds, there is the muck brought in the house on your boots from the loft floor. No matter how many times you scrape your boots before going in there is always that one time you forget, and furthermore I would sooner have a bad trap in the early races and be in time to win if they had trapped than be half an hour behind with a good one.

Some of the best pigeons I have had have been bad trappers on the short races, but when the distance is reached they are my best trappers. Like the good red hen 'Shy Lass'. She would have won double the 1st prizes if she had dropped from the early races. I have topped the Fed with a pigeon a minute or so behind her. But when 'Shy Lass' had been on the wing for seven or eight hours she would almost go straight through the door into her nest box.

If I had done what I had heard a lot of fanciers say about bad trappers and got rid of her I would have been minus one of the best hens I ever had from the distance. Like many others I do not like to get beaten on the trap. It is hard enough to win, but as I have said they will trap when the right time comes.

I am sure that it is in the early races where bad trapping occurs and it is not just because of heavy feeding. All the years I have kept pigeons I always get keyed up on the first two or three races as regards to a bad trap. It does not matter how they have come in the year before, pigeons in the early races get excited, especially the yearlings, and these are the ones who should be sitting on the second pair of eggs and should be coming into form.

My wife knew me like a book, so did the kids when they were young. She used to say: "Be careful, your father's had a bad race," and not another word was heard out of them. I have often had several good birds out when I have shut the loft up. I did not have many pigeon conversations with my wife — not after some 50 years; and I might have been sitting watching television, she would look at me and say: "You are some good pigeons missing."

"How do you know" I would say, and all she would say was "I can tell!" I always put on a good stone in weight during the winter, but soon lost it when racing starts. No man worries over his birds more than I do when it is a bad race; birds will come through fire and water, but not through mist and fog. No matter how good they are, this is something they cannot beat.

Those left behind

When good birds are out at night, I am sure they have tried too hard to get home, become confused and gone wrong. I am always up early the next morning to see these in, as it is no fault of theirs they have had to spend a night out due to a bad liberation. Not only do I get up early to see them in, but also to get them back on their eggs, as birds will recover more quickly this way. This is always the most worrying part of bad races, it mucks up all the loft with the mates that are left at home.

When the birds go away Thursday, I let the mates that I have kept behind sit all day Friday. On the Friday night I put them in a basket with drinker on; cocks and hens separately. I make sure I close up the nest box. As I have said many times I have no nest fronts on my nest boxes, but I have several pieces of wire that slide in front of them. This is most essential, as birds like to come home and find their nests as they left them. The eggs, I put under other pairs that are sitting with their mates.

I have sometimes had four or five eggs under a pair, not because I want them to hatch, for, as I have said, I only want to rear one pair of eggs from my race birds, but because it is much easier to get birds back on warm eggs than stone-cold ones.

Getting bird back on eggs sometimes needs a lot of patience. I have stood in my loft over half an hour but have always succeeded. I have found the best way when they come home is to get the cock or hen ready to come out of the basket and put the warm eggs back. Then drive the one that has just arrived home into the nest box, making sure he is standing over the eggs before putting back the mate who has been in the basket. As the eggs are warm, they will think the other has been sitting all the time they have been basketed up. Getting birds back on eggs after a hard fly is most essential, as I have said birds will recover more quickly as they will get some rest after a hard fly, and when birds are sitting they are resting.

Some Feds are not giving the birds a chance to get fit. 220 miles in early May, this may have been all right years ago, but as I have said the seasons are changing and recent years have been no exception. The cold north-east winds are no good for getting pigeons fit. There is something money cannot buy when getting pigeons fit, and it also puts the needle in the pigeon's brain pointing the right way. That is the sun. The plants in the garden the other year were three weeks behind, therefore pigeons are the same. To expect pigeons to fly 200-250 miles in early or middle May without having a sun on their backs is the cause of bad races, and the convoyers get the blame. Every year more blackbirds are building their nests in sheds or outer buildings, why? Because the trees are not covered owing to the late spring. I said all these things at the right place (in Fed meetings) and got nowhere. Sure, I do not know it all, but one thing I do know: pigeons will stand big jumps when fit but not before.

Big Jumps

I am sure my theories paid off a few years ago. My best pigeons had not been in the race basket for three weeks, and had missed heavy losses among the fanciers that had sent. This was not sorting the wheat from the chaff, as I have heard them say. Every fancier I have spoken to has dropped a good pigeon and as I have said before one bad race can ruin a pigeon for the rest of the season, and you will not have time to get them right for the Classics. The pigeons even in my own loft, that win these races get no medals from me. If the same type of race was flown from the same distance the following week, you would get a different pigeon first and perhaps lose the winner of the previous week. When it is a true race you get the pigeons you expect, and only with this type of race do you sort the wheat from the chaff.

I find that I am getting politically minded! I am not a pigeon politician, but at all times when I write or talk like one it is in the interest of the sport in general!

Every year when I wean my second round of youngsters I keep a watchful eye on the first round that have been weaned off earlier. I remember seeing a young blue cock running round and feeding those I had just weaned off. That happened to me three years earlier. I bred a nice young blue chequer cock, I could not fault him when I weaned him. Several weeks later after I had put some more youngsters in the loft to wean, I noticed he was always tucked up. I made up my mind he had gone wrong and the dustbin was the place for him. As this was the weekend and the bins are not emptied till the Monday I thought I would wait until then.

On the Sunday I was doing something in the garden when I heard youngsters squeaking. Thinking one was being knocked about in the young bird loft I went in, and there he was feeding two at a time. This was taking too much out of him, and why he was always looking sorry for himself. So I collected up all the second-round youngsters and put them in another loft for a week; when I put them back there was no more trouble as they could now get up in the aviary and were picking up well. I find it is only when the youngsters are on the floor that this sort of thing happens.

The pigeon in question was my good blue chequer cock 'The Laird', the best pigeon I had on the road. He won from 78 miles and 502 miles. Some fanciers think a long-distance pigeon cannot win from the short races. When pigeons are fit they will win from any distance. Some of the best pigeons I have had all won from the short races before winning from the distance. One of my favourites 'The Combine Cock' won the Fed from 55 miles; 1st Fed from Thurso 502 miles the same year. These are the ones I want.

Winning Short and Long

No workingman can afford to keep two teams of pigeons, one for the short races and one for the long, and as I have said I like to win them all! To me it is ridiculous to concentrate only on long-distance racing, I get as much pleasure from the short or middle races when they win. Little fish are sweet, they help to pay the corn bill. True some pigeons do not show until you start to reach the distance. I could name a dozen in my own loft who never won a prize until they got to 250 miles, why, I do not know, especially when the nestmates won from all distances. Perhaps they took longer to come into form.

I know some fanciers think birds do not come into form until they have cast their first flight. This is not altogether true, I have won some good races with birds that have not cast. When pigeons are fit they will win in all conditions.

I am sure a lot of pigeons do not keep their form because they are fed wrongly on arrival from a race. I do not care if they have only been on the wing for two hours, my birds never get the main diet on arrival. I always close the hopper which contains peas and maize so they cannot get at it, and feed them all seeds which are easy to digest: groats, rice, lentils and plenty of baked bread crushed up not too big, then the next day a good feed of maize.

There is nothing better than maize the day after a race to bring birds round. This is something I always take notice of, the way pigeons pull round after a hard fly, or have been out for a day or so from a bad race. Those that take four or five days to pull round will never make long-distance pigeons; they have not got the constitution. But I am sure you can help them to pull round if fed on the right grains on arrival, plus plenty of greenstuff, watercress etc. This is the only time I put anything in my drinker, two tablespoons of honey in a cup then pour hot water to dissolve it. Hot not boiling, as otherwise it will take all the vitamins out, then add this to your drinker.

Pigeons seem to crave for salt when a bit low down, and I am sure eat more minerals than are good for them when in this state, so I put half a cup of table salt in their bath and leave it there all day. You will notice they will keep going and having a drink. They forget the minerals. Common table salt in the bath is a wonderful antiseptic after a race, and keeps them free from any infection they may have picked up in the basket.

As I have said, I have no time for sick pigeons, but most of the minor ailments of pigeons can be treated with the same things we use ourselves. A pecked or ulcerated eye; warm boracic crystals or Golden Eye ointment, or better still penicillin eye ointment, all have done the trick.

I am always pleased when the youngsters are away from my race birds. I know we should not race our birds when rearing youngsters in the early races, especially if you want the youngsters, but we must get them in some time before they get too far up the road. I always leave one at home to look after them, and try to switch the youngsters around.

If I have a pair that have been kept at home or my stock birds have only one youngster I put one of the youngsters from the pair of which

I have sent one to a race with it. I make sure it is the same colour, blues and blue chequers are OK, but do not put in a red or a mealy, the parents will know the difference. Don't think pigeons are colour-blind like some fanciers do.

I have had a pair of my stock birds feeding three youngsters while one of the parents is away. I am talking of birds going away Friday for a Saturday liberation. Getting back to the point in question it is very distressing to see a pigeon coming back after a hard fly and go straight and feed them. That is why I like to see all the youngsters out of the racing loft when we start racing in earnest.

The old birds need all the energy they can get for the Classic races, feeding two youngsters saps their strength. It is birds wanted for the long races that will be in this condition, for mine must be paired up late as regards the wing moult. That is why for years the birds I wanted for the long races I only let rear one youngster from each nest. This way you can send one to a race on alternate weeks and the parents and youngsters show no ill effects. But letting them rear two youngsters can cause a problem during the racing season.

Chapter IX
What Price Pedigree?

What price pedigree? That's the start of this chapter. Before I give you my views on the subject, I would like to tell you a strange but true story. Out shopping one day a woman picked up a young street pigeon that had fallen from its nest from under a railway bridge, feeling sorry for it she took it home. Knowing one of her neighbours kept pigeons she took it to him, after hand feeding it for a few days he put it with his own youngsters, where it soon learnt to pick up and was soon on the wing with them. The next year as he was short of a hen he paired it to one of his cocks, and rung one of the youngsters from them. As a yearling this youngster took several positions in minor club races. The next year it was 10th Open LNR Combine Thurso 503 miles, a very hard race. During this time the hen in question had changed hands and was given to another novice who after a time did not like it, so killed it. After the performance from Thurso, the fancier went round to get her back. It was then he was told what had happened to her.

I have never wasted too much time in studying pedigrees. When taking in a cross I have selected the pigeon I wanted from a reputable loft and have then been told its parentage. I have always let my own judgement be my guide, and can honestly say I have never failed. Good healthy pigeons are what I require, not long bits of paper.

I remember a good friend of mine who I thought would make a good fancier. He purchased eight youngsters from someone else with pedigrees as long as his arm. I remember him coming to me when he received pigeons and pedigrees and all he kept on about was the pedigrees. I then asked him what the pigeons were like, but he had not got the same praise for them as he had for their pedigrees. From this, I took a different view of him becoming a successful fancier. To cut the story short, all he had left at the end of the year was the eight pieces of paper. Badly reared youngsters with pedigrees of any length will not help them to win races. I've always made the race basket my pedigree since first coming into the sport, and still do.

Take the dam of 'Red Admiral' for instance. She was unrung. The year I was looking for a cross for 'Blue Lad' I went and saw a good friend of mine with whom I have been exchanging pigeons for years. His brother had spent a lot of money on the best he could get. One he did not get was the old 'Plum' mealy cock I wrote about earlier.

He handed me several hens; some that his brother had paid good money for and taking particular note of the eye of all of them. I told him they were not the eye I was looking for.

Then I spotted a red hen in one of the squares and I asked him to catch her, taking her outside the loft in the sunlight to look at her head and eye. This was the one I was looking for, and he told me his brother had bred her and forgot to ring her, saying she was out of this and that, but I was not interested in how she was bred. Again I preferred letting my judgement be a better guide than pedigrees. She not only bred 'Red Admiral', but several good pigeons for its owner and was named the 'Unrung Hen'.

Strays and Street Pigeons

Getting back to the street pigeons, I keep hoping to read that the RPRA is going to do something about getting them cleared. They are a menace to fanciers living in large towns, within 100 yards of me there are at least 700, the youngsters drop on your loft with roup, canker and every known disease under the sun. They worry the life out of you, especially when you do your utmost to keep your pigeons healthy. The same applies to strays, I know they are a headache to any Union, but they are also a headache to the fancier. Any stray I treat with the utmost suspicion; I think they must have something wrong with them else they would get home. I let them enter the loft and when they do they are immediately caught. Then they are put in the garden shed with food and water and after a couple of days are released.

One of my club member's birds was flying well; then for two weeks he was almost at the bottom of the race sheet. As he was flying a large team of pigeons I thought this was strange. The next week while ringing birds for a race we noticed several birds that had dirty wattles and were suffering from a cold. They were immediately taken out of the race basket to check the ring numbers. They all belonged to this same fancier, who then agreed they were not fit to go to the race.

Surely, when one's birds are flying well they don't go off form like that unless something is wrong. At all times when this happens one must look round and find the trouble. Birds with dirty wattles should not be sent to a race, not only for the sake of other fanciers' birds that will come into contact with them, but are not in a fit condition to find their way home and will go down. I think all clubs should appoint someone of repute to see that this sort of thing does not happen. Colds among pigeons are the most common disease. Damp, overcrowded and badly ventilated lofts are the main cause of the trouble, plus, as I have mentioned, coming in contact with street or stray birds that are affected. Another cause is when birds have been lost for a week or so, and are flown down, so that their resistance is at its lowest; old birds that are rearing their third round of youngsters, and are well in the moult; when the youngsters are badly in the body moult with the strain of growing the new feathers; these are all times the resistance to the cold is at its lowest.

This can easily be noticed, sneezing birds who when cleaning themselves have the down feathers stuck on their beaks, where the mucus has run down and made the beak sticky. You will see streaks of yellow, like dabs of yellow paint, on the box perches, this will tell you which bird is affected. I have no time for sick pigeons, meaning those with the major illnesses; canker, going light, ornithosis, leg or wing weakness. I am not like a lot of fanciers I know, who have a medicine chest full of pigeon cures and preventives. If pigeons are kept fit and healthy there is no need for any of this and the man who is everlastingly curing his pigeons of this and that will never win races.

But no one can help getting a cold, and my cure then was Avisol made by May & Baker although I do not know if it is still available. I could guarantee birds would suffer no ill effect after and would clear up within a week, and the wattles will come back to the right colour. This is not white as most fanciers think when a pigeon is fit, but a pale pink as though they were transparent. These birds are in tiptop condition and are hard to beat.

End of Season Care

When old bird racing is over, I spend most of my time with the youngsters but do not forget to give the old birds all the food they require, plus a handful of linseed every other day in pots to help the moult. I stop giving them an open loft. They are kept quiet, let out once a week to have a bath and pick about the garden at the spinach. Watch the weeds that grow in between your spinach plants. Bell-bindweed for instance — this they will pick at and it will make them very ill. Dandelions and chickweed are good for them. I remember many years ago having all the green foods that I have mentioned over a time when writing, analysed, and chickweed came out on top with much more percentage of iron but it's not always available. The time

CHAMPION MICK
Winner seventeen 1st prizes and four times 1st London NR Fed, 1st Fed, 4th London NR Combine Open Berwick (4,668 birds); 1st LCC, 102nd Open LNRC Fraserburgh (3,875 birds).

to pick it is when it has gone to seed, and hung up in the loft the birds will pick at the seeds — it's a wonderful tonic.

On my frequent visits to Belgium, the land of the pigeons, I have seen barrow-loads sold in the market square in Brussels. The market place is like Club Row of London used to be, mainly pigeons and other birds, where you could then buy some beautifully-reared youngsters for about £7 a pair, plus ring cards and pedigree of sire and dam. But that's another story which I hope to deal with another time.

Pigeons must at all times have a certain amount of iron to keep them fit and full of energy. Some fanciers use Parrish's Chemical Food, some iron drops in the water, but I like mine to get it the same way as birds do in the wild state — from green food. That's all a wood pigeon lives on throughout the winter, plus acorns, I have shot wood pigeons and when cleaning the crop have found it full of clover. I counted 26 whole acorns in the crop of another and there is no hardier bird than the wood pigeon. That is why I don't mind the clover growing on my own lawn. I have seen my birds eat it readily. Mine get a good feed of acorns during the winter months, these my grandchildren collect, doing what their mothers did when they were young for their pocket money!

One can read over and over again the same things written by different fanciers, clean food and grit, plenty of greenstuffs and baths and so on, but it's the little things that count; these make the good fancier. I am sure you can tell a good fancier by the chips in his basket. When birds are brought into the clubhouse I always empty into the dustbin the wood chips that I have used the day before and always put in a couple of handfuls of clean chips for the next toss. If birds have been sick or brought back their food this will go sour in the basket. Birds will readily eat it the next time you put them in and this will put them off form. Never take birds to a race with nothing in your basket, like some fanciers do. Don't forget, you are not going to race ring them. Don't use sawdust, this can blow in the pigeon's eye, and can make the difference between a good toss and a bad one.

Never dust your pigeons with insecticide the day of basketing for a race. This can make the difference between first and last, as the insecticide can make them dopey. The time to do this is after a race and definitely not before. I cannot remember the last time I had to dust the pigeons after weaning. Lice won't live on healthy youngsters, but thrive on the sick ones.

Caution with Late Breds

Don't overcrowd your loft with unknown late breds, the price of corn is too high. Only breed a few from the very best and only the best of these should be kept. Late breds make excellent stock pigeons, for they are reared in the best of English weather with long hours of daylight. I am then asked what I term as a late bred. Those hatched from middle of June onwards, as they are too late to get in the first races.

I have stopped breeding late breds to put on the road next year. It does not matter how much training you give them in the first year, they never beat the early ones as yearlings, for by the time they get on the wing the early ones have stopped running out. This is why they make a lot of mistakes as yearlings. Youngsters that run out for three hours will learn more than all the training you care to give them; they cover miles of ground and late bred youngsters won't do much running on their own.

I can hear someone saying: "What about old so-and-so's late bred cock?". True, you get the exceptions, but what a good pigeon it would have been if it had been bred earlier; getting that bit of extra education from running; that's my way of thinking. Of course, when one loses all his youngsters in a flyaway, a few youngsters next year won't hurt, but don't expect big things from them the first year. Take them up to about 150 miles, put them in all the short races you can, plus a couple of 50-mile tosses each week. Always pick a good day for them and don't let them meet a hard race or toss, else that's the finish of them.

I have never had to use this kind of patience as I have only had one flyaway in my life, and that was back 1958. This was a lovely morning, not a cloud in the sky; they had been flying round home for a few days about half an hour each morning, this time 12 of them went up like dots and away. I got one back late that evening and one the next morning and never saw the others. The one I got back the same day was my 'Blue Girl', winner of the Young Bird Combine, the one I got back the next morning won the London Championship Club the week after, so I began to think the others were not good enough.

I remember speaking about this to my good friend, George Lovell of Hatfield, one of the best fanciers to fly in the London NR Combine, before he unfortunately had to give up the sport owing to ill health. He never let youngsters out at this age when the sun and the moon

SILVER QUEEN.
Dam of fifteen 1st prize winners when paired to Champion Mick. Granddaughter to Scottish Lass.

were out in the sky together. Strangely enough that's just how it was when I let them out that morning and you never hear of them, or get them reported. Many good fanciers say they fly higher and higher and get into space, who knows? But one thing I am sure of — a lot of flyaways are caused by letting youngsters out with an empty crop. They overfly themselves and have not got the strength to get back.

Hopper feeding is the way to make youngsters fly and go running, they eat just enough to satisfy themselves in the early hours of the morning, and will go far enough till their tummies start gnawing, telling them just when to turn back. Hopper-fed pigeons have not got a full crop and don't put on too much weight, therefore flying them is a joy. Directly I pull the house curtains and my youngsters see me they fly on to the door I let them out by, and are waiting for a fly. That's how youngsters should be, they are gone within a minute out of sight.

Changes in Homing

I flew pigeons at my old loft for about 50 years and ever since I can remember my youngsters have always gone running straight out the west. It's something I have tried to work out for years, but have never found the answer. Half a mile away in that direction is Alexandra Palace with its television mast. I sometimes think that they make this their landmark and homing device, but one will never know the secret of how pigeons find their way home. Many have gone into this, but none have come up with an answer that satisfied me, but I like to think the brain of a pigeon is like a radio, picking up wavebands and the sound getting stronger as they near home, with the sound and landmarks to help them pick the shortest way home. When pigeons are 100% fit their batteries are fully charged, but when birds are tired and not quite right their batteries are run down and the homing signals are much weaker and only rest will recharge them.

One thing I am sure of, pigeons are not homing like they did years ago. In those days I could shut my loft up with them all home within half an hour. Not today, you get three or four together, then wait 15 minutes before you get two or three more and this goes on the rest of the day. Don't tell me it is clashing, we have had the sky to ourselves and the same thing happens. No, it's the man-made electricity and jet aircraft that are in the sky today. They are breaking up the wave beam that pigeons home by. Last but no means least the entry fee is far too cheap for Classic racing. A few shillings more would see the best a man could send and not those that are being given their last chance before they are put in the dustbin. So much for pigeon politics; I am only guided by my thoughts when writing.

Youngsters go through different stages according to the weather. When it's dry and warm you say to yourself what a nice team of youngsters. But get a few damp cold days and they don't look the same. You can't get out of the loft fast enough. You wonder what's gone wrong and you cannot see a good one among them. So never let birds sit outside the loft on this type of day. Birds love the rain when it's warm and it helps to straighten the bent flights that you may not have noticed. In the same way these birds are always ready for a bath.

Bent flights are caused by the gaps in your box perches or by fighting. The flight should immediately be put in the steam from a boiling kettle. Make sure there are no gaps at the top of your box perch squares, these can also cause them to be pulled out or broken and will lessen their chances of winning. Never pull a broken flight. I have known some fanciers who have cut the flight about two inches from the stem of the quill and have drilled down the quill and glued in another flight. I don't know if any of them ever won with this but I would sooner mine take their chance in a race with a flight broken.

Chapter X
Eyesign in Breeding

When I started to write my postbag became far greater than I could cope with. One particular fancier wrote and said he had had a pair of pigeons together for three years, and had bred three pairs of youngsters from them each year, yet they had never bred a hen, always two cocks in the nest. I am afraid all the years I have kept pigeons I can never remember this happening in my loft. I very seldom have two pigeons paired together all that length of time, unless they are breeding good ones. I am always trying to breed something better and will switch the cock to another hen, likewise the hen to another cock.

I have had some pairs that always breed two cocks or two hens in a nest, never a cock and a hen. My good blue cock 'Mick' as far back as I can remember only once bred a cock and a hen in the same nest, always two hens or two cocks. As I have said for many years, when this happens one must be a good one, and invariably is, but it's not always the one that breeds the winners, it's the nestmate. I remember some years ago breeding two hens in the nest. One was the first silver blue I had bred, coming from the first cross of the Ameel blood. The other was a blue which was my good blue hen 'The Jet' winner of thirteen 1st prizes and £500 from all distances, good money in those years.

The Silver Blue was picked up twice walking the streets flown out, once in Spalding from a 104-mile race. Three weeks later I put her back on the road, the same distance, again she was picked up, this time in the Birmingham area. After a lot of bother with the fancier who picked her up I got her back. This hen was one of the best violet eye hens I have ever bred. Her sire was 1140, a son of 'Plum', and her pedigree was far greater than her homing ability. But I stuck to my opinion on eyesign for breeding and to cut a long story short, she bred me four blue cocks to win over £2,000 in Combine races: 'Sure Return', 'The Combine Cock', 'Record Breaker', and my good 'Blue Cock 4408', jumped straight in Berwick 300 miles, 4th Open LNR Combine when four years old.

Her sister 'The Jet' never bred a pigeon worth mentioning. This is where eyesign plays the most important part in pigeon racing, knowing those that are no good on the road but will breed you winners. I'm sure many a good producer has been killed for this reason. How many champions breed winners? There are more good pigeons bred by chance mating than those we put together on paper. I know of a pigeon that was sold for a good price; the parents were used as feeders for the first nest, the second nest the fancier rung one from them; this was the pigeon in question. It won right through from Lerwick! When fanciers say they will only breed from those that win on the road, what a big mistake they are making. True, one likes to put winners to winners, and you can hold your own with such matings. But not in my case. Some of my best producers have never been in the basket. To me this is the only way to keep at the top, knowing the ones to keep for stock.

I have studied eyesign as far back as 1938 and can honestly say I have never read a book on the subject. When I say never, I did have an eyesign book sent to me by a South African by the name of Fell. The first page said that eyes with egg-shaped pupils are degenerate and should not be bred from. After reading this, as the book cost me nothing, I threw it straight on the fire, as the eye in question, egg shape from two o'clock to eight o'clock have been my best breeders and racers. This was the same eye as the Blue Cock 1140, 1st Fed, vel 795 yards, the next week 3rd Open LNR Fed Morpeth, 256 miles, young birds, also a winner from Lerwick.

I am sure the true effect cannot be put down on paper for the knowledge of other fanciers, but one thing I am sure of, those that pooh-pooh it would not let me take away birds from them on eyesign alone! I'm not as clever as some fanciers who think they can tell the ones that will win and those that will not. All types of eyes win races, but only the few with the right eye breed the winners. True there are some eye colours that win more than others. These are the eyes with the winning gene showing in the eyes of their grandparents. Where this happens one must always try to reproduce this type of eye; that will help you keep at the top. After the war most of my pigeons had the violet eye and were winning out of their turn. From these I bred one of the best racing eyes, what I term the red, white and blue: bluish-grey around the pupil going into a shade of white, then the remaining iris, a wine colour. Whenever I bred these they were Fed winners, and when I visited a good fancier's loft and saw this type of eye, nine times out of ten they were the birds that were doing the winning. Then the eyes of my family changed from the violets to the browns, and the orange red with a full green circle, and these occasionally bred a violet.

BLUE CHEQUER COCK, NU69L19765.
Winner seven 1st prizes, 2nd Hexham 5-Bird Specialist Club beaten by Flash, only two birds on the day. A good pigeon that never raced again owing to hitting wires.

Violet Eyes and Pupils

I remember many years ago, before the war, the Editor's father, the late Major Osman, with two great fanciers from the North whose names I cannot remember, giving a lecture on pigeons in general and also showing the slow-motion film of pigeon flight belonging to The Racing Pigeon. I have mentioned that I learned a lot from this film, as regards the importance of the upward movement of a pigeon's wing. The question of eyesign came up and of all the fanciers I have talked to, who are like me — eye-minded — one of them put it in a nutshell. He said he had never done any good with pigeons with big pupils. But he added don't go home and kill those with such, I don't think anyone talking on the subject could have put it more plainly. I remember the 'Plum' mealy cock about to tread his hen outside the loft. The pupil went as small as a pin's head.

One does not need a watchmaker's glass to see a good eye; this is something I have never used. I have spotted birds with good eyes outside fancier's lofts. Depth of colour is the main factor when selecting eyes for breeding, and when taken to the sunlight the pupil should retract to the minimum like the 'Plum' mealy cock. It should be made up of three different shades of colour; the majority around the outside of the eye. The pupil should at all times be out of shape, meaning not a proper circle. Let me put it this way. Give a child a pencil and ask it to draw a circle, that's the shape I mean.

The violet eye is wanted by most fanciers but these can be very misleading. There are some that are 'dormant' and will not breed a thing. The violet again must have three different shades; the first being a bluish-grey around the pupil then going to a reddish violet, similar to the red, white and blue eye but with more depth of colour. The best of all are those with a green circle; these only come with the orange and light brown eyes, and when taken in the sunlight will occasionally flash the shade of a bluish-green.

What with some fanciers saying they are colour-blind, I may be wasting my time, for compared to the enjoyment and ease I have usually in writing I find writing about eyesign the most difficult. Don't think I am trying to back out on my word. I said I would prove to those who pooh-pooh it there is something in eyesign. As I have said, let me come in your loft and take away those that fit in with my eyesign theory, and see how sick your loft will look. Give some fanciers a pigeon and the first thing they will look at is the eye, even those that say there is nothing in the theory. Give them a washed-out eye and hear their remarks. No one likes light-eyed pigeons because firstly it does not go with the make-up of a pigeon's head and secondly, they don't know what else to look for. True, there are good eyes in some lofts that never win but give them to the right man and see the difference.

Selecting the Eyes

My old friend Jimmy Dolby of Acton, at the time when he was new in the sport, asked me if I would have a look at his pigeons, as he had spent quite a lot of money and was wondering why he was doing no good. I could see he was very keen and most downhearted so I went over expecting to find something wrong with his loft or with the food he was using. But the loft was perfect and so was the corn, then when I saw the pigeons, I knew they were a bad lot. This I told him, and said that even I could not win with them. He never spoke for a few minutes and I said I was sorry if I'd hurt his feelings, but that's what he asked me over for. Then he said he had just bought the last eight from a a fancier who had packed up. Straight away I said these are different. Going through all of them I picked out a blue cock and a red hen. "These will breed you winners in every nest," I said. Not only did they do just that, but they also bred the Western Home Counties Combine winner from Berwick, and the 2nd Open Combine from Thurso.

Ask the older fanciers in the Dalston area what I did with 12 pigeons that were in pens for me to give my views on eyesign. After making my remarks on each individual pigeon, not one fancier could contradict what I had said about them. Again I was asked to pair two of them up on eyesign, this I did and at the end of the year I was told they had bred three 1st prize-winners and a 2nd Open London Championship Club. One Saturday morning I received a letter from a friend of mine with a money order, asking me to pick him out two hens from a dispersal sale of Vandeveldes which he also kept. They were being sold by a fancier a few miles from me, and were advertised as £5 each. So over I went with a pal. When he saw me he was most surprised, "Don't tell me you have come to buy pigeons." "Yes," I said, "a friend wants me to pick him out two hens." After handing me several I said to the friend who was with me, in a quiet voice, "These have not got what I want."

Then he gave me a blue chequer hen which had what I was looking for. I said: "I'll take a chance whether my friend shoots me or not. He wanted two hens for this money. I'll give you the lot for this one." To this he replied: "Do you know what she is, she is the LNR Combine winner from Berwick, over 9,000 birds." He also would not part with two daughters from her that had the same eye, I shook hands and bid him good-day with the money still in my pocket. I picked out the sire of Jimmy Biss's LNR Combine winner 'Mackeson' from Fraserburgh on eyesign from the loft of George Lovell of Hatfield. I have been quoted as being bombastic when I have said it's 60% the fancier and 40% the pigeon, no doubt after reading this you will think me more so, but I'm only trying to prove my words. A great statesman wrote his autobiography and when asked why it was never published, he said after reading it, it made him seem bombastic. But if written by someone else for him, it would not have read the same. I have never known a good pigeon fancier when talking of his ways and his pigeons of not giving the impression he is bragging. It's a pity a few more good fanciers do not talk more freely, perhaps we would learn more and when one is talking of his good pigeons he is only proud of what they have done for him, and if he was not he would not be a true pigeon man!

As for putting the highest percentage on the fancier in front of the pigeon I can put this in a nutshell. I loaned a friend of mine a good hen to pair to a good cock he had. He raced five youngsters from them without scoring with either of them, the next year I had one from the same pair, 1st Fed, the next week 2nd Open LCC, again proving my point.

The Winter Months

Some fanciers leave their birds together all winter and part them a few weeks before they pair up. I have always parted my pigeons the last week in August and my hens are not let out till they are brought round to the racing loft in January. Then they are let out twice a day when the weather is right. True, they have a job to get airborne with the certain amount of fat they have put on during the winter months, but within a week they are running out.

I like to keep my birds quiet while going through the moult and at this stage I stop all the small seeds I have mentioned, just a bit of linseed every other day. Small seed is only required to keep your

BLUE WHITE FLIGHTED COCK, NU70L32207.
Winner of five 1st prizes, 1st Morpeth, 48th Open LNR Combine (5,195 birds). This cock was picked up with a damaged wing and was not raced again. Son of Champion Mick and The Mosaic Hen.

birds under control and to get them in on race days; given all the year round makes it not so tasty when required and will keep them too much on their toes. I like my birds to go through a form of hibernation during the winter months. Just plain good sound corn and rest, it does pigeons good to put on a bit of weight, it all helps to give energy to grow the new feathers and have a perfect moult. It can easily be flown off before breeding starts.

When the holiday-makers start leaving seaside resorts here and on the Continent we start to see more pigeons return that were lost. Perhaps an old favourite that was dropped at a certain race point will return now that he is not getting fed by holiday-makers with their peanuts and crisps. Those that home with their race ring still intact can sometimes be of great value, and potential champions. I could name a dozen birds of my own loft that have returned after months away. They have never made another mistake and have earned their corn each year. The good 'Mealy Hen' was lost at the YB Combine. She came home in February the next year, still with her rubber on. Her flights were a bit worn, so I gave her three short races then stopped her. She had a perfect moult that year and became one of the best hens I ever flew. 'The Laird' was lost as a yearling from a very bad race after winning four races on a trot. He returned the following March still with a rubber on that was nearly perished. He is the best racing cock I had in the loft at that time, a winner from 78 miles and 502 miles. When you breed from this type you will have no fear of their youngsters entering the first loft when tired and hungry. One reads many times, over and over again, the conditions some pigeons race for home, eggs, youngsters, driving, and so on, but they pull out their best for the love of a good master and I'm sure that's why they don't go on into other lofts.

Chapter XI
Young Bird Racing

When you first come into the sport there never seems to be enough races, but I'm afraid as you get older things are different. You are glad when racing finishes, at least that's how I am nowadays. As I have said, getting to the top is much easier than staying there, and to do this one must eat, sleep and dream pigeons as the saying goes. The only time I could relax is when the season ends, and my mind wanders to other sports where someone else does all the work and worry.

I've had many discussions regarding young bird racing and my opinions differ from most. When you read young birds should not be raced, what a lot of tommy rot. I like young bird racing as much as old birds. And I like to see how the youngsters from a new cross, or from a new mating perform. When you hear fanciers say that they don't take any notice of young bird results, again they are talking out of the back of their heads. Most of my best pigeons have won or performed well as youngsters, and no fanciers' youngsters flew harder than mine when I was really going for it. Twice a week, 200 miles on the Wednesday and 200 miles at the weekend, and all my youngsters have to fly the race programme, apart from those I have stopped for stock purposes.

My two best yearlings flew 200 miles three times and 256 miles twice and were the only two birds on the day at the latter as young birds. Only good sound healthy youngsters can stand this type of schooling. No youngster should show any signs of distress after having been on the wing for six to seven hours from the distance mentioned. If they do, they will never survive the old bird programme or make a name for you. Those who say they are not interested in young bird racing show a different view when they have a good one. There is not a true fancier today who, like me, does not want to win them all. True some fanciers fly better young birds than others but only because of the type kept.

I have spoken many times of the size and type of my pigeons. If they were just the opposite, big and corky, I'm afraid that with all my knowledge and my past performances I could not win with them, and when you think of the moulting condition we race youngsters under no wonder the type mentioned above does not win. As I wrote in my first chapter I won six out of eight young bird races in my first year, and since then have made notes of the winners as regards the state of the moult.

I have topped the Fed with only six tail feathers, the rest moulted out. In fact I have topped the Fed with youngsters in all kinds of condition except when heavy in the body moult, meaning with the bars or cover feathers missing. This is a big handicap. When youngsters are in this condition, not only is it a big strain on the constitution, but with the bars missing the downward thrust of the wing has not got the full force behind it, because the air penetrates through the wing. The bars and cover feathers counteract this. Take a thin piece of cardboard and flap it downwards, put another piece on top and see the difference. There's your answer.

Take the wing condition as an instance. We would never send our old birds to a race in the condition we ask our young birds to fly. But they home! Why? Because the frame of the pigeon is not fully grown until they are at least two years old. While the body is developing so are the flight feathers in length and width to carry the body. As I have said before, I am not a wing faddist, especially with youngsters. The further up on the wing the better the body moult, which I have said is the difference between first and an also ran.

I remember many years ago breeding a nice young chequer cock, he was on the clock six times out of seven winning four 1sts, twice 1st Fed. I kept looking at the state of his wing hearing so much of the old wives' tale of what a bad condition it was when the 9th flight was out. Naturally he was my Number One candidate for the Combine. Four days before the race he cast his 9th flight. You can tell how disappointed I was. I only pooled him up to 5s (25p), but he was only beaten by a decimal for 1st Open LNR Combine at the weekend. It taught me a lesson. A few years later I put all my money in the Young Bird Combine on six youngsters that had cast their 9th flight and won over £400, also 1st & 2nd LNR 2B Championship Club. When youngsters have cast their 9th flight the danger point is there alright, but it is the body moult. I once verified a 4th Open Young Bird Combine winner with the last flight missing, only nine full flights. When I pointed this out to the fancier he said he did not worry about the wing as long as they could fly, but I would never dream of sending a youngster in that state of the wing and never have.

Young Birds and Clashing

It's only when the young bird racing starts that we keep hearing the word clashing either at the race point or *en route*. To me the latter is less important for reasons I have written about before. I can honestly say I don't lose many youngsters when racing starts.

You read "don't let the youngsters out on Saturdays when the old birds are racing". For years when I have timed in, out go my youngsters, for the sole purpose of getting them mixed up with the late race birds. This does not apply to the novice, but when you are established you can take chances. Two or three times a week my youngsters, while running out, clash with other youngsters that are doing the same, and come back in twos and threes all day long. That's why clashing has never worried me and I would never fancy my chances with a team of youngsters who had only flown round and round the house top.

You also read how to start training young birds two miles, four miles and so on. I often wonder where these youngsters will get their knowledge from. The only youngsters that need this type of training are those who have never gone out of sight of the loft and they will never beat those that have roamed the countryside while running, and have got split up and had to find their way home. No one knows the amount of ground they cover, but one thing I do know is, they are learning the shortest way home. When they get split up they will have the confidence to fly on their own, apart from getting muscled up for when the time comes to stay on the wing.

So before we start putting it down to clashing when we get a bad young bird race, and it invariably happens in the second or third one, give a thought to the following. Have they been properly schooled? Have they spent a few nights in the race pannier with the drinkers on? The first year we went over to road transport using crates, I made one almost the same as the Fed ones and made my youngsters stay several nights in it. Have they been properly trained? I start mine off at ten miles the same place every year, near some radio pylons on a direct line of flight, making this their breaking point. Three times at this point then straight into 35 miles as many times as I can before the first race. All this is done ten days prior to the first race. Never start training youngsters too early. If they are still running directly you start training they will stop running, and while running out will learn more than all the training you can give them.

Like my old birds I never single them up; I put them up in twos or threes, this way they will race one another home. What a boring task singling up can be and I am sure they learn nothing of value. Never train on a full crop. If your birds are hand fed they must be got in straight on arrival with your trapping mixture. Another thought before blaming clashing, have they been properly reared and are they fit and well to meet up with any trouble that may occur? It makes you think when you know of the same fanciers every year who get cleared out before they get to the fourth race. Surely you cannot put it down to clashing because when their birds are put in the race pannier and when they are race rung they go beserk and upset those that have been properly schooled.

Too many fanciers rely on the other fanciers' birds to bring theirs home. This I have heard them say many times, and these are the first ones to moan when they don't get them. I only wish there was some way to prove that youngsters have been properly trained before they could be entered for the first race. Too many untrained birds at the race point spoil the others and too many are allowed to reach the stage of ringing that should have been put down from some defect or other, let alone reach the first race. I don't care if they are from my Number One pair, if they are not right I cull them. When you hear fanciers say they will make up there is no such thing in pigeon racing! Youngsters must be right as soon as they leave the egg, else they are doomed. Too many youngsters suffer from malnutrition, some fanciers buy 28 pounds of corn for the old birds and expect to rear a team of youngsters on the same amount. If all youngsters ran out like mine and went through the same schooling and treatment the word clashing would be forgotten.

Young Bird Moulting

As I have said, I have made a close study of the moult of youngsters, because at this stage I have found some of my best pigeons. I don't like youngsters to fall to pieces all at once. These have been stuck in the moult and have never made a champion. Youngsters should go through a steady moult if all's well. Those that clean up before you even noticed they were in the body moult, these have always made my best old birds. They are the ones that start to moult their secondaries when seven flights up.

When we turned over to road transport my youngsters held their flights longer than any year I can remember. Youngsters hatched March 8 from my stock birds are generally on their 9th flight by the second week in September. That year they had just cast their 8th and had completed the body moult. They were at least two weeks behind other years in respect of the wing. I was most puzzled, but they flew again very well. As this was the first year my club had gone by road transport and were travelling in crates, I thought that could be the answer. We all know how hot it used to be in the railway wagons when fully loaded with our youngsters and this could cause them to moult more freely. Also I had not seen the same amount of body feathers in the loft after a race as there were when we went by rail. This could also be the reason why youngsters were taking a week or so longer to clean up, but it made no difference in their future career on the road.

As you will realise my young bird system differs from most. I can hear the young fancier say: "It's all right for him, he is established". True, but when I first came into the sport on my own, I made the race basket my pedigree, fed them well, flew them hard. That has always been my motto, and I have held my own ever since. I don't believe in mollycoddling pigeons. So while you still have them fit and well, send them. Only stop them when you think they have earned a place for retirement.

All my yearlings go right through to almost 400 miles. One year I timed eight in from that distance to win money, four were yearlings including my second bird, over 11½ hours on the wing. Not one of them showed any signs of distress. In fact the next day they could have gone back and would have again been hard to beat. Four hundred miles is the ideal distance for yearlings and from these you can look for future 500-milers the next year.

I remember my good friend, from whom I borrowed the dam of 'Red Admiral', had a good mealy hen half-sister to 'Red Admiral' from the same sire 'Blue Lad'. This hen flew every race to Thurso which she won three years running, with a good position in the Combine each time. Someone told him how to win the Combine with her. This was by not racing her so hard and jumping her from Berwick to Thurso. This he did and he never got her on the day. No, this hen loved racing and was kept fit by doing so.

They kept telling me I would lose the old 'Plum' mealy cock, by keeping on sending him and my words then were: "I never stop a winning pigeon". He broke the record in the Wood Green Club by winning eight successive 1sts. I remember the last one he won I had taken money in side bets, not because I wanted to, but my hand was being forced. What a relief when he dropped in the pouring rain, 1st Club, 2nd LNR Combine Berwick, the next week 1st Fed, 7th Open Yearling Berwick. That's what I call flying them hard, and good pigeons thrive on it.

The chances of pigeons surviving on the road today are far less than they ever were with the amount of television aerials there are today. These damage or kill our birds more than all the wires and cables ever did. So while you still have them, fly them. Don't put them on the shelf for next year as the first race you give them they could hit an aerial and you have kept them all the year for nothing.

I don't like youngsters that keep coming back from the opposite direction; those that have flown over. These are followers and have not got a mind of their own. When they get to the distance as yearlings after relying on other brains to help them home, their own instinct becomes dormant and they are lost as soon as they get in the wrong batch. I never encourage my young hens to lay like some do by pairing them to old cocks. True they fly well, but I want mine for future racing and not for just one season. You can hold your own with a healthy team of youngsters, and other systems that do not stop their growth or shorten their racing career. You all know what I mean; young hens to old cocks or vice versa, but I will deal with this in my next chapter.

Chapter XII
Pairing youngsters for racing

I never encourage my young hens to lay, knowing many fanciers do by pairing them to old cocks. True they fly well, but I want them for future racing, and not for just one season, especially those that are paired up before they have completed the body moult. I am now speaking of past experience, these I have found fly invariably the same again as yearlings. As soon as a young hen starts sitting the moult will slow down or stop completely, and after several weeks they will fall to pieces as though they have been put through a plucking machine, and will come almost completely bare, and are now in the same condition as those that have been stuck in the moult, something I do not like, which I have already written about. True, if I see a young hen that has been playing up to a young cock, for several weeks, become eggy I give them a nest bowl but only those that have completed the body moult and have one or two flights to go before the last flight. These will not only win as young birds but will win again as yearlings.

As I have said, I do my utmost to stop my youngsters from racing under these conditions. One can invariably hold one's own with a healthy team of youngsters that have been properly schooled and trained, and other systems can be used that do not affect their growth and shorten their racing careers. You all know them, young cocks to old hens and vice-versa. This should only be done when there is something worth winning. Let us deal with the first one; old cocks to young hens. There is something about old cocks that all the young hens go for, these are the cocks to use. I find this out by letting some old cocks out with some young hens. When you have found the right cock, the afternoon before they go away let all the other cocks out of the racing loft. I am now talking of our final Classic when my birds have been parted for a fornight, leaving only the one the young hens have chosen in the loft. Then I put the three or four hens in with him for about one hour before basketing. The older the cock the better. They seem to pick a cock about five or six years old knowing it is not spiteful. The year I won the Combine with 'Blue Girl' I had five hens with a nine-year-old cock. They were almost fighting over him. I timed the five in under three minutes; they were all in the first 20 Open Combine. All were nicely cleaned up and over the body moult. They were on the 9th flight and had flown very consistently previously, but just needed that little bit of urge at the right time to win.

Young cocks to old hens, again must be over the body moult. These I get suited to an old hen three or four days prior to the Combine by letting all the youngsters out of the loft, leaving only the two or three young cocks I want inside, then I bring round three or four old hens that have been parted, to put with them.

These are mainly the cocks that have commandeered the only three nest boxes I have in the young bird loft for this purpose. When I see that they have taken a hen into their nest box I collect the hen up and put her back in the loft she came from. The day of basketing for a race I again let the youngsters out except the same cocks, then I bring round the hens for an hour before I basket them up.

As I have already said, one can hold one's own with a healthy team of youngsters that have been running out and have been properly schooled, but one must know which bird to put the money on. This brings me to a point I must stress, never pool beyond your means, there is not such a thing as a certainty in the sport of kings. Take the great horse 'Nijinsky', he found one better, but I'm sure he was over the top with the amount of work he had had. The same applies to pigeons — there can always be one better, there is nothing more depressing than to lay out more money than you can afford on pools when they meet a bad race, or one that does not suit the birds you have sent. I remember when I won the Fed in the Berwick race with my mealy cock 'The Plum'. My daughter, who at the time was nine years old, knowing the chance he had of winning, wanted to open her money box and pool him more. I said then that he could get beaten.

Parting for the Autumn

Frequently we get wonderful weather in late September, just right for those who, like me, had parted their birds for the moult. I like to see the birds well cleaned up before the bad weather sets in. Some fanciers still think like I used to years ago, that the longer you leave them together, thus holding back the moult, the better the feather for next year's racing. I have found it makes no difference if you part your birds like I do late August or the first week in September, or leave them together another five or six weeks. The longer you leave them together, the more likely that when parting them, they fall to pieces and become almost bare. This may be all right in a warm loft, but not in a loft like mine that was well ventilated and with the damp cold nights that time of year. By late October the birds are feeling the cold and they don't like it. It can cause them to have a bad moult, therefore the feathers are not as good on these birds as those that have finished the moult before the bad weather sets in. When you have a family of pigeons with good quality of feather, "a touch like velvet", there is no need to worry about the feathers standing the wear and tear of next season's racing. I once said that I never let my hand be the guidance of a good pigeon, meaning bone structure, size, depth of keel etc. Perhaps that is not altogether true, I have never handled a good team that were coarse and dry in feather. With all the oily seeds you can give them, you will never achieve the good feather coating mentioned above if the texture of feather is not there to start with. I never handle my pigeons when they are heavy in the body moult.

I invariably have visitors in the winter months but as I have said, birds need quietness. Apart from being the worst time of year to see a man's pigeons I'm sure a good fancier would not let you handle them if you did pay him a visit. Some fanciers leave their pigeons together till a month before breeding and then hold their own when the racing season starts, but I like all my hens in one loft, where I can weigh them up much easier, long before the breeding season starts. I am always "boxing" my birds around, to try to breed something better than I did the year before. When you have a winning family of pigeons that are moulded to your ways of flying and management, it is the hardest job in the world to find another family that will beat them, not because they are no good, but because they don't answer to your ways and means of management. How often do we hear of fanciers who have had a good winning family of pigeons packing up, restarting again, but never flying the same as they did in the past?

One must always try to do better than the year before, to keep at the top. Never discard your own family for some new strain that has come into fashion like some do. There are good and bad in all families of pigeons. Take the Cattrysse family as an example. My good friend Reg Barker made them known in England with some outstanding performances. My old friend Les Davenport put up that wonderful performance in 1969-70 in the Pau National with the same family, but one invariably hears fanciers say that they have tried the Cattrysse and found them useless. A poor fancier can get the strain a bad name. If a poor fancier were to take over my loft of pigeons he would say the same after flying them for one year. It's the man that makes the pigeon, not the pigeon that makes the man.

I don't think we give enough credit to the fancier who has not got a loft in his own garden. Those, I mean, who keep their lofts on plots some distance from the house. We all know the amount of times we go up to the loft for some reason or other when the breeding and racing season is upon us. There is always something we have forgotten to do or give them. It is quite easy when you have the birds at your beck and call, but not so easy when you have to travel to look after them. These "plot" fanciers do this because they love pigeons, so full marks to those who fly under these conditions. The man with his fabulous loft often cannot hold a candle to the backyard fancier. There is something money cannot buy, stock sense and good management.

The working man can often only afford to buy his corn each week, thus taking a chance whether he gets the same quality. I like to purchase enough peas and maize to last me through the whole old bird racing making sure I have a bit over. Then I purchase enough to take me through the young bird racing. This way I am sure of having the same quality grain. Many a loft has gone off form through changing the corn during racing, but not everyone can afford to do this by purchasing corn in bulk. They often have not the room to store it, which is most important.

North or South Road

Having flown the North Road for over 50 years, the last years at the old loft, I put a few youngsters on the South Road as a matter of interest. I am sure more fanciers North of the Thames would fly South if the facilities for training were much easier. I think that when you train pigeons you must get them there and back in the shortest time to be of any benefit. One often hears of fanciers commenting on the hardest route that pigeons fly. They say the North route is downhill. We all know the mist that forms in the Channel when birds are liberated from France and Spain, but we North Roaders get the same in Scotland. Sometimes directly after the birds are liberated down comes the mist, or failing this, they run into it an hour after they are liberated. Quite different to the South Roaders, they do have a chance to get their line of flight. I leave those who think the North Road is downhill to think again. There's nothing I like better than to get in a free-for-all with pigeon talk.

One argument I will not be drawn into is: Which is the hardest route pigeons fly? I said many years ago, and I say it again now, 500 miles is 500 miles, on any route. Beyond that distance in my opinion is out of reach of the average pigeon. In racing terms the longer races should not be called races but marathons. I think too much of my pigeons to chuck them away at Barcelona or the Faroes, and even Lerwick from where I've had my glory. I remember having a word with a friend of mine who lives in the Midlands about why we in London don't fly Lerwick. That was before my Fed put it on their race programme some years ago. He forgets the distance from Thurso to London is further than he flies from Lerwick. When the LNR Fed first flew Lerwick they had over 300 birds. After four years barely half that number went, with only one bird in race time, proving my point that it is beyond their reach. Lerwick to London is one of two things — a smash or a blow home, and for me I can do without it, if I can get my pigeons to win from 500 miles I am well satisfied.

I wrote about some of our old favourites returning when they were not getting fed by the holidaymakers, bearing in mind a good blue cock I lost at Thurso. A week after writing I received a card from the RPRA reporting him in Holland in a small town called Den Helder. In fact, the last town in the North of Holland where it stretches out into the North Sea, approximately 160 miles due east of Cromer. This would be the first sight of land a pigeon would have when blown out on a strong west wind such as there was from Thurso. After making arrangements to get him back, my patience got the better of me so I decided to go and get him. On arrival at Ostend at 4.30 in the morning, we motored up to Den Helder, over 300 miles, in not good flying weather, visibility down to a mile. We eventually arrived there just before dark, where I met Mr J Nebbeeling who had reported him.

Knowing the hospitality of the Dutch and Belgian fanciers it was not surprising we were soon made at home. He told me he had reported over 100 pigeons belonging to English and Scottish fanciers. I now quote his own words: "Scotland have plenty of money, England no", meaning that Scottish fanciers always want their pigeons back. Not so the English! He seemed very disappointed with some English fanciers not wanting their birds back. He also said a week after the Thurso race hundreds of pigeons were flying back and forth around Den Helder. Like the blue cock of mine, many were picked up exhausted. He then told me that the blue cock had been sent to the town of Utrecht where it was awaiting shipment to England. I was a bit disappointed not being able to bring it back with me, but the information I received from Mr Nebbeeling was worth noting. My record still holds, the Bakers won't go in. I now know where the Thurso birds went on July 12 and I am sure there are still many over there on the Continent.

So far those who have their favourites missing, I hope you have the luck to get them reported like I did. Although the race controller was much criticised for liberating in a strong west wind, this will always happen when the wind is in that direction and as strong as gale force. Birds will go with it when liberated on the coastline. If birds were brought inland ten to 15 miles, I am sure they would have a chance to steady up and think twice before going over the water.

Mr Nebbeeling also showed me stacks of letters from all over the British Isles, from those who lost their dear ones in the last war. His job was superintendent of the war graves. He told me some letters brought tears to his eyes. Knowing how sincere he was in the short time I was with him, I am sure the graves were well looked after.

Judging in Ireland

I don't want to finish this chapter on such a sad note, one must always look to the future, and that is the shows of which I have in the past had the honour of placing the cards. Cleanliness and condition are of No 1 importance. I would never take a pigeon out of the pen that was dirty and out of condition. Bad fret marks will peg a pigeon back to the minimum, don't try pulling these out thinking they will moult a good feather, the judge will notice this immediately. I am speaking of racing classes, I don't like show classes, they have not got the same effect as a good lot of racing pigeons. They give me the impression they would drop on the first house when the going got hard. Still they are show pigeons. Some years ago I had the honour to judge in Ireland, that class of pigeons still sticks in my mind, I can see some of them now, the best class of pigeons I ever went through. It's a pity there were almost 120. I only wished I had more cards to put up, that's how a judge should feel when he has gone through a racing class. This proved to me what I already knew, the Irish are a good lot of fanciers. I had to handle every pigeon and it gave me a couple of hours of pleasure. I have had classes just the opposite, where you cannot find enough pigeons for the cards, and have been most disappointed. If you give the cards to pigeons you would take away you cannot go far wrong so No 1 rule is cleanliness and condition, then the judge will have to take them out to fault them.

Chapter XIII
Waiting and Watching

"Keep your eyes and ears open, your tongue still and you will go through life the right way". These were the words my mother used to say when I was a boy. How right she was, especially when I became a pigeon fancier, and where there were pigeons I was not far away. What a treat it was when some of the fanciers used to let me see and handle their pigeons! Some were as big as ducks and some just the opposite in another fancier's loft. But they all had their share of winning. I knew then that as long as the heart was in the right place the size made no difference to the art of winning races. In those days they all fed and raced their pigeons differently with an equal amount of success. That's why I've said many times there is no book of rules to the pigeon game. I am sure they would not have changed their winning ways just because someone was doing better.

Several members of my club have built their lofts identical to my own, they use the same training points. They are only doing what I did many years ago. If you do get the opportunity to see a good fancier time in, don't ask too many questions while he is waiting for his birds, as his mind is on one thing — the pigeons that he is waiting for. It's surprising what runs through your mind while waiting for your birds; who will be the first or from which way will they come. Much can be learned on the arrival, even from the short races, the way they trap on the conditions you have sent them. Look always for those that show signs of distress.

My late arrivals get the same treatment as the early ones, as the next week they could be my first pigeon. Nothing is more disturbing than for someone to keep asking questions when you are expecting them; that's why when I was a boy and kept quiet I was always invited a second time to see a good fancier time in.

The off season is the time to get all the necessary jobs done to the loft, never leave until tomorrow what can be done today. I'm sure this phrase could have been meant for pigeon fanciers. First a coat of paint on the outside. When I built my old loft the back was lined with hardboard. At the back of the nest boxes I fixed it rough side out to soak up the creosote. The young bird loft I fixed it smooth side out which gives it a good finish with the white emulsion which I've always used since it first came on the market. I always pick a good drying day for the job. After letting the birds out, I wash it down with boiling water and it is completely dry within the hour ready for the birds to come back in. I've never painted the inside of my loft, solely because I have not got another compartment to put them in for two or three days while the paint is drying. I find emulsion is quite adequate for making the loft clean and keeping them free from lice and red mite, especially if you add half a bottle of Duramitex and it is done two or three times a year. That's why I have never had to dust my pigeons with insecticide after they have been weaned.

Don't leave your nest bowls outside so that the snow and frost will crack them. I remember up to a few years ago when we could go to our local potteries and pick out the misfits for a few pence each. Today they are many times that price but I am sure there is no substitute for earthenware nest bowls, unless you have time to make boxes 10in square by 4in deep, then creosoted and left to dry before using. These have made excellent substitutes when I have run short of nest bowls.

Nest Box Fighting

Transferring the young cocks over to the old bird loft can sometimes be a problem unless this is done as soon as the young bird season ends, as I like to get them well acquainted to their nest box before the breeding starts. This all helps to stop fighting at the wrong time. I make a habit of going up to the loft late at night and putting them in the empty box I have chosen for them. After several nights of this practice I find I have no more trouble. When you keep pigeons you never seem to have enough lofts for your requirements. I always wish I had had three compartments, one especially for the young cocks to race to as yearlings and then when they are two years old transfer them over to the old bird racing loft. This would not only solve the problem but the headache it can cause with some old cocks that have decided they want two or three boxes but I have always managed, with a little patience.

I cannot remember the last time I went through my pigeons without finding one or two with fret marks across the flights. I've heard fanciers say it's a sign of weakness. What a lot of poppycock! It all depends on the state of the flight when sent to a race. If the flight had been more up in the wing with the hard fly or a night out it would not be visible. You will notice the majority of the frets are within one inch up the wing, proving my point — the flight had not completely broken through. Some birds fret through mental strain, even from the short races; those that have got off their line and have been lost for several hours and have worked back to get home. I am sure that's why a lot of birds fail at the distance, not through physical fitness but mental strain which saps up energy far more than flying. This weakens the homing instinct and they are lost, and those that don't regain it are never seen again unless reported. With those that home under these circumstances after several weeks, it is because something has clicked in the brain and they are straight home. As I have said, these will often pay the corn bill and may never suffer from this form of amnesia again.

Some of the unfortunate feeble-minded people are as strong as an ox, but when it comes to using their brain they show their weakness; the same applies to the racing pigeon. Don't think those frets will lessen the birds' chances for next year. I have heard fanciers say they won't do any good till they have moulted them out. I was beaten once by a pigeon that had the worst fret marks I have ever seen across the flights. It had been out for several weeks as a young bird and learnt from its mistakes. Those that are fretted on the third flight will have to go to the long race with it. I would sooner mine go with the fretted third flight than moulted out and on its fourth.

We all have our likes and dislikes with regard to colour. I have never known a fancier who does not like blues or blue chequers. Something like 80%, the majority that go to the race are of such colour. The other colours are in the minority of winning. I have always said a good pigeon cannot be a bad colour. From my own family I bred two white cocks; they all knew in London of their performances from the long races. I could name several others that were the fancier's best pigeon over several years. I remember in the bad winter of 1962-3 seeing a white sparrow in my garden and I don't remember seeing it show any effects to the hard times they were having. In fact, when it came to the grub-stakes it was most aggressive and always had its fill before the others could get near it. If whites were in the majority to the blues and blue chequers sent to the races those who think the colour is a sign of weakness would think differently.

Keeping the loft clean

Why some fanciers want to use V-perches or saddle perches instead of a nice row of box squares I will never know. To me pigeons don't look right on them and I am sure they are much easier to catch or pick up from a box square than the others. How nice, too, the birds look when the boxes have been whitened. Perhaps it is to save scraping the droppings from them each morning, as we know if you powder the saddle perches with limestone powder the droppings will fall off. Scraping out the loft is something I have never tried to shirk, and at each weekend when I do my floor and shake a handful of lime over it, I always look back and admire how nice it looks.

Some fanciers put two or three inches of sand down and sieve it each week. This a friend used to do till I told him the birds would pick and eat it because of the salt from their droppings. He stopped it straightaway and was soon doing much better with his birds. When the time comes that I cannot scrape my perches every morning and the floors at the weekend I will pack up. I can hear someone say what about old so-and-so, he does not scrape out for weeks, or they are on deep litter. Yes, I know and have seen it, but I could not keep my pigeons under these conditions. I often wonder how they win and find their pool pigeons. Birds look better in a nice clean loft and I can assure you while I have been scraping out I have often noticed something to my advantage for the following week's race.

Earlier on I said I would write about my visits to Belgium and the lofts I have visited. I was more than pleased to see in the Pictorial the lofts of George Goossens in an issue some time back. I don't know what it is but when you have lived pigeons as long as I have, there is something about a good fancier as soon as you see them, and G Goossens gave me that impression straightaway. When he picked up his pigeons to show me I knew he was a born fancier. Although he kept many colours, they looked as though they had come from the same mould, not too big, just the right size. No wonder the pigeon that put up such a good performance in South Africa was a Goossens-Baker, as they are an identical type with my own. They are housed in perfect surroundings well above the ground. Like most of the lofts in Belgium, even the backyard lofts are well away from rising damp, and most of them that are able to do so have some form of heating under the loft, proving what I have written about pigeon lofts.

They must be kept dry to keep them fit. You could travel for miles around Belgium without noticing pigeon lofts unless the birds were out. They are not built like the English with a wire or dowel front, the backyard lofts are completely boxed in. Only on the front will you see several panes of glass and a drop board where the birds enter.

Some Belgians Visited

In Maurice Delbar's loft there were radiators on either side, also some form of heating directly under the loft which is in use during the winter months. What a wonderful fancier! The birds were a shade bigger than the Goossens and of only two or three colours, mostly blues or blue chequers with the occasional white flight, built more for the distance. Although my Flemish is not all that good, with the bit I picked up during the war I always got through to them, and the hospitality of the Belgians is out of this world. I am sure there is a bond among pigeon fanciers second to none, no matter what language you speak. That's why it is such a wonderful sport.

The amount of good friends I have made over the years makes me proud I am a pigeon man. Like the unforgettable visit we made on the De Baere Bros of Nokere. Again the birds were housed in perfect surroundings over some old stables with a boiler underneath that supplied the two cottages with hot water and kept the loft dry during the winter. We visited them quite unexpectedly but within ten minutes you'd think we had known them all our lives. The birds that had been raced on the Natural or Widowhood system were now rearing a late youngster, but one could see that they were of the highest quality and not only by their past performances from all distances. They were of a similar type to all the Belgian lofts I had the pleasure of visiting, always with a good head and eye, a head which shows the full amount of character, when one is looking for good pigeons, something better than lengthy pedigrees.

The lofts that gave me most pleasure were the lofts of N & R de Scheemaecker of Antwerp, where 25,000 pairs of all the known strains are kept. Each strain is housed separately in a 100-foot loft, with the same size wire aviary on the front. To me this was a feast and a must if you ever visit the land of the pigeons. The loftman, or foreman as they call him, was then a likeable old gentleman called Jan Willox. As this was on a Sunday, the large lounge and bar was filled with pigeon men and women, all waiting for their number to be drawn from a large drum. De Scheemaecker sells everything appertaining to the pigeon sport and when you have spent a certain amount of money your ticket is placed in the drum, and if your number is drawn out you can have a pair of squeakers from any of the strains kept, giving everyone a chance of obtaining a strain of their choice. As Sunday was Jan Willox's business time we made arrangements to see him the next day with the object of purchasing some youngsters.

This was late February, but I can tell you there were plenty of youngsters about the farm, as the Belgians cannot get their rings early enough as they have to fly the old birds for the amount of any-age races organised.

When we arrived the next morning the youngsters were all in separate boxes with a lid at the top, where the strain was pinned on. Myself, I had not gone with the same object as my friends as to purchasing youngsters, but I could not help falling in love with a little chequer hen of the Havenith blood. I asked him the price — 600 francs, equivalent to £5 — I brought her home. As a young bird she won several positions and the next year 70th Open LNR Combine Berwick, 5,444 birds, and was timed in from Thurso the same year. She was one of the very few yearlings I have sent to Thurso and then only because she was barren. I lost her a fortnight later in a smash from Perth. Not only did she win much more than I paid for her, but again she had proved my judgement right. Don't forget we can all be a good winner, but not all can be a good loser. When you have fulfilled your boyhood ambition you can look back with pride and know that all was not in vain.

Chapter XIV
Is Widowhood Unbeatable?

Let's start the chapter with Widowhood versus the Natural system. No doubt many fanciers like myself will try to do better than they did the previous year. This is the only new year resolution I make, as I have broken the others far too often. This is the only way to be successful, but novices who put their birds on the Widowhood system are trying to run before they can walk.

If you cannot win with your birds sitting or on the Natural system you won't get them to win on any other. To hear fanciers talk you would think that the Widowhood system was unbeatable and essential to win races. Only the other week I was talking to a fancier who showed me some cocks he was going to try on the Widowhood system. If he was to pay more attention to the way he looked after them — barley all over the floor among the droppings — I am sure he would do much better and would not need to worry about any other systems.

I know some of the best fanciers in England hardly know what the word Widowhood means, they depend entirely on the Natural system. If they flew their pigeons against the crack pigeon men of Belgium like those in England they would know that it's not as good as it's made out to be, or of such importance, especially on this side of the Channel, where the glory comes from races of 400-500 miles.

The Belgians get the same amount of glory or honour, call it what you like, from only half the above distances; where there is a lot of money to be won and where no doubt Widowhood comes into its own is as an aid to quick trapping. I have flown a semi-Widowhood system for years. The number of 1st prizes some of my cocks have won tells you that, but I have won just as many 1sts with birds sitting the right time.

I have had three or four cocks semi-Widowhood, only to be beaten with hens from my own loft which have been sitting eight days. Don't think I am against the system. On the contrary, to be successful with pigeons you must try them all, but first get them to win Naturally. Only the skilled fanciers can get and prepare their birds right for the Classic races on the Natural system, and I am sure, will always hold their own against the Widowhood cocks. I often wonder how the Continental fanciers would do on this side of the Channel.

I still say 500 miles is 500 miles in any direction, but the country the birds fly over makes a big difference. The cold fronts and the

SILVER PRINCESS
Winner four 1st prizes and 1st London Championship Club, 35th Open LNR Combine YB Northallerton (5,274 birds). Grand-daughter Red Admiral and Scottish Lass.

English Channel are no help to them, and when you know the Belgians give a position for every four birds sent to a race, compared to one in approximately 40 birds sent in English races, the competition must be harder here to get cards, but nevertheless they are good pigeon men. I only wish they felt the same way about us. They always give me the impression we are a lot of novices compared to them, but I am sure it's because too many fanciers, if you can call them fanciers, have gone there with their cheque books hoping to buy success and buy anything with a Belgian ring on.

I repeat, let some of the crack fanciers this side of the Channel fly against them in their own country and I am sure they would tune them up and the Continentals would soon alter their views. But for every good fancier born in England, there are twice as many on the Continent. I have seen boys of seven to eight handle pigeons like experts.

During the winter months fanciers tend to forget part of the main diet of the pigeon and that's grit. Pigeons need this at all times to help digest the food, to get the full amount of vitamins etc from whatever corn you use. A little every day is far better than a potful left in the loft for a week. At this time of the year it will get damp, and dust from the birds and loft will stick to it and do more harm than good. Why buy the best corn you can afford only to leave one of the cheapest things we give our pigeons in the loft for days? The same applies to mineral salts. This I have stopped giving my birds during the winter months as it absorbs moisture.

Rest in the Winter

I have never been one for exercising my birds too frequently in the winter. My hens have been shut up since September and my cocks have only been let out once a fortnight. After Christmas my hens are put in the young bird loft and let out every nice day. They will be right when I want them for pairing. My system has never failed me yet, and I know they have not worn their new flights out flying around home.

I remember a few years ago a fancier phoned me to say he looked like winning the London Section from Lerwick with one of my blue hens, velocity over 1600. We were at Fraserburgh the same day doing around 1300. I asked him how he had got on from there. "No good", he said, "blow home". I was most confused. There he was doing

1600 ypm from Lerwick, and he called it a blow home from Fraserburgh with a velocity far less, but that's typical pigeon talk.

I do know I have never timed in a "stumer" in this type of race. Getting back to the mild weather, I don't think anyone likes the cold, but I am sure it does not hurt pigeons. They look better when the weather is dry and cold, and it helps them to settle down and become more content than the mild weather in the middle of winter. Birds get too excited and think of one thing, going to nest. One has to keep a watchful eye to prevent hens mating with one another.

I have never cut my pigeons' rations in winter, they are hopper fed all the year round, and when I see two hens that are always in a corner I part them for a week or so, but have always made a note, as in the past these have always been my best racers. I have never worried about them putting on that extra bit of weight which will be most beneficial to them when the cold weather starts. By hopper feeding I am sure they are all getting the same amount, as some birds are slow eaters, and nine times out of ten these are your best pigeons. I'm sure if I had fed my pigeons I would overfeed them in kindness.

A fancier from Telford, Salop, once told me that my eyesign theory differed from others regarding the egg-shape or distorted pupil of my best breeders, but my writing deals with my methods and my birds. With all due respect to others' ways and means of denoting eyesign I have not read one of them. In fact, over the past 50 years and more I have not read much pigeon literature. When one is gifted with the full amount of stock sense, call it what you like, I find some other ideas hard to absorb.

Over the years I have proved many times how wrong some of the articles that have been written have been. I have my own set ways with pigeons, and no one in the world will change them. I have learnt from my successes and not from mistakes. I have made many a mistake and have learned nothing. Money makes money, success makes success, and to do this one must be most observant and think like a pigeon. I can go back years and know how a pigeon won a big race, meaning the condition sent. I don't have to look it up. If I did I would not find it as I never had to keep records on how they won.

Chapter XV
Brain comes first

I would like to deal with brain versus brawn, or muscle versus the homing ability. I am going to stick my neck out and give full marks to the latter. Every bird was born to fly. The racing pigeon was bred to fly long journeys. No one can tell me they were bred entirely for muscle, if they did we would never have got them to fly the distances they do. No one knows how many times winning pigeons stop in the long races when they do 500 miles on the day and have been on the wing for 15 hours. We all hope and think they have been on the wing without having a breather but one will never know. But I am sure the brainy pigeon will only fly the shortest way home, unless blown off course by a strong side wind, and the amount of muscle given by Nature will always see it home.

Not so those which lack the full amount of homing ability, although having an abundance of muscle for staying on the wing. These will fly round and round the race point trying to pick up the loft direction. Those which after flying miles off the line of flight eventually do arrive looking as fresh as paint only to be well beaten by the brainy ones, which know directly the strings are cut where they are going, and have no fear of mental strain sapping up their energy. Only when we breed one with the combination of both do we get the champion. But how many of these do we breed? If I could breed one every five years I would be well satisfied. No one wants to breed from pigeons which are always two hours behind the winners, and home looking well. I have tried pairing these to the winners, trying to find the combination to make a champion, but have never succeeded.

I now keep my brainy ones together ignoring physique and have held my own doing so. Everyone can find the muscle of a pigeon when in the hand, and nine times out of ten the pigeon will prove them wrong when it comes to winning. I have had many pigeons in the past that in my own mind have not handled like 400- or 500-mile winners, but they have proved me wrong. When given to other fanciers to look at I can tell by the look on their faces they think the same. When told of their performances from the distances, they have been most surprised. The number of very small hens which have won long races, I am sure is equal in proportion to all other size pigeons sent. I have never worried about the size of my hens for racing.

Take the long distance runner, he is like a human pin, and would never stand a chance as Mr Universe, but my money would be on him when it came to a marathon. I must admit I have not the same views as regards small cocks. These I don't like and have found they have failed beyond 300 miles and are what I call "sports". I try to breed my cocks of a medium size, and after all these years have succeeded in moulding a family to my liking. Only then have you created your own strain. Give me the light framed pigeon, buoyant and a good head, full of brains. These, with one flick of the wing, are up from the garden. You can have all the others. You will notice these, when exercising round home, hardly move their wings to keep with the others who seem to be making hard work of it.

Youngsters on the loft

When one is presented with a youngster, before crossing it with your own family, we all like to see what it is made of before putting it on the road, and compare its ability against our own. To me this creates a lot of interest, but I am sure one can find out the exceptionally good ones without racing them, even when they go running out. They are those which always come back with the first batch, or, when they get split up, have to work back on their own and don't go too far adrift. A few years ago I liked the mating of two pigeons a friend of mine had put together, and he promised me one from the second nest. When it was ready he phoned me to say it was 24 days old, and would I come and pick it up. Unfortunately I was down with the 'flu, and was unable to pick it up for another week. When I got there it was on the top of his loft. He assured me it was the first time it had been out. This is something which I don't like. Youngsters should not see the outside of the loft if bred for someone else and should be taken to their new home before 24 days old. They settle down much easier and will not fret.

After keeping it in for three days, knowing I might have trouble with it when my own youngsters had come back from running, I opened the top and let it come out. As soon as it did I could tell it was not happy with the surroundings. It was looking more curiously at the objects in the garden and I could see it was going off. After about 20 minutes it did, straight to where it was born. About four hours later I phoned my friend but it was not there but I had a feeling I would get it back, and to my surprise it was on the loft the next

SURE RETURN, NURP49SS9963.
Winner of ten 1st prizes. Second Open London NR Combine Fraserburgh (5,407 birds). Grandson of Plum.

morning. He did this for almost a week, trying to find where he was born but eventually gave up trying. This told me all I wanted to know without flogging him on the road. He was one of the best stock cocks I had, and in every nest he bred a 1st prize-winner, including two 1sts Fed on the North Road and one on the South.

I also loaned him to my friend who used him for one round of eggs which proved to be two 1st prize-winners in another Fed! He became one of my favourites, not because he bred winners, I have several good cocks which do the same each year, but because he proved my judgement right and this gives me great satisfaction. After all these years I am not slipping.

I complete the necessary jobs that need doing before the breeding season starts. The loft has been painted outside, the inside with white emulsion, the nest bowls have all been boiled, scrubbed and stacked away ready for when wanted. The fence and wire round the garden has all been repaired to stop the cats. My next job is the training baskets. These will all be thoroughly washed down with the hose, and when dry will be given a coat of varnish and any lining that is torn will be renewed as I don't want them coming home limping after a toss, because legs have caught up in the webbing, which sometimes happens. It's surprising the state of baskets some fanciers take their birds for a toss in. They have not seen a drop of water or varnish for years. No wonder they hold the others up on the race sheet, and give the sport a bad name.

When you talk of pigeons to the average person, they think straightaway of those around Trafalgar Square. No wonder, when they see these baskets and bad impressions are created. There are those fanciers who don't want to help their pigeons to win by spending half an hour washing and varnishing the baskets. Think of the sport in general and do your utmost to give the boost it needs to make it equal to other sports in Britain.

Chapter XVI
Breeding to Type

There are many sides to the sport of pigeon racing that give one as much pleasure as the racing season. Foremost is the breeding season. Some fanciers have a knack of pairing two pigeons together and breeding winners each year, and this comes from success and observation over the years. When you have a successful loft, the winning genes are so plentiful I'm sure if you let your pigeons pick their own mates you would still hold your own, providing you have a basic strain as your family, and not as most fanciers do when they start, go to different lofts to obtain their stock.

Just because you have purchased winners, or sons and daughters from them, does not mean you will produce the winning genes. On the contrary they will be different shapes and sizes, and last but by no means least, a different type. The latter, to me, is the main factor when mating two pigeons together to produce the goods. A big cock to a small hen, or vice versa, would be useless to produce the perfectly balanced pigeon that to my mind is most essential to win races. I would never pair a long-casted pigeon to a short one. Likeness in head shape and length are the main things when I mate my pigeons, using the deepest colour eyes, full of richness, never the same two colour eyes paired together for breeding winners. The violets to the red, white and blue, or the red pearl eye, which comes down from the violets thus using the same basic colour, the dark brown eyes to the orange red eye, again from the same basic colour coming down from the dark brown eyes. But by letting them select their own mates; although holding your own, you would soon lose the basic type you started with.

It is this base which one is always looking for when one matches a pair of pigeons together, the pleasure and achievement would not be as great as from using one's own judgement and from the result of study through the winter months before mating. Even if in your own mind you have matched a pair of pigeons to perfection, it does not mean to say the first nest will be what you were looking for. There are so many genes in a pair of pigeons and you don't always get the right ones. I'm sure one cannot condemn a pair by only taking one nest from them, but must give them every chance to prove their worth as breeders, and that your judgement is correct. One must take at least three nests from them, leaving them together for the whole breeding season. I am sure that the longer they are together the truer the type.

I sometimes had to wait till the final nest to get the type of youngster I had been expecting. Surely when you match a pair of pigeons together one always visualises the type they should breed, but as I have said, it is not always so. I stopped the mother of 'The Laird' at three years old. She proved herself by winning several prizes from the distance, but the type she was breeding was the main factor and she was breeding winners. I knew I could win with them before I ever raced them, as they were typical of the type I had held my own with over the past years. I am sure there are many fanciers who could kick themselves for trying to burn the candle at both ends. Good producers, even though they are excellent racers, are not worth sacrificing, as the amount of money they would win on the road is nothing compared with the amount their progeny would win and how much they would help to keep you at the top.

The same applies to pigeons that win from the same distance each year, using the old phrase, horses for courses. I am now talking of races of 300-400-miles. Don't be too eager to push them on. Wait till they become sluggish or cunning and you have had the best from them at that distance then push them on. These 300-400-milers are not little fish, they are the middle cut of the salmon which is very tasty and to me is the ideal distance to call a race. Nine times out of ten these races are won by the most consistent pigeons previously over the course, and by taking them further too early in life you take the edge off them, and make them plodders.

Too many 300-400-milers have been chucked away by chasing average trophies. I would like back some of the good middle-distance winners that I have lost at a final Classic, which can sometimes be called the graveyard for this type of pigeon. Not because of the distance do they fail, but it is the type of race they meet, weather etc, from that extra 100 miles, and when the true racing pigeon runs up against a heavy belt of rain, trying its hardest to get home it goes round it. Those that are some distance behind, reach the rain area when it has eased off or cleared up, keeping them on their line and leaving the early ones to fly miles off their course and forcing them to have a night out.

The Older Breeders

I have said before I am only guided by my thoughts when writing. Let's deal with old pairs, those that bred winners the first time of asking. These are always your first mating, hoping that once again they will do the same, but one forgets the older they get the slower the progeny. I never rely on old pigeons to breed my racing team. By doing so you are inclined to overlook the others but one is always tempted. I am sure a lot of producers fail when they get past eight years old, because they are allowed to rear their own young. I have a pair of pigeons that had bred winners for ten years mated together, but after the age of seven years, I've always transferred their eggs under good two year old feeders.

When pigeons get past eight they do not make the full amount of soft food that is required to give youngsters a good start in life. Not only do they fail to do this, but old pigeons are beginning to look for an easy way to feed their young, and instead of giving them wholesome grain, they start to feed them with too much water, which is much easier to regurgitate than peas. They become useless as parents like the wet feeders. Some pigeons deteriorate far quicker than others. Those that show a sign of rheumatism or have come through a bad moult at the age mentioned, I would not breed from. Although in the past they may have bred winners each year, they are over the top and past their best. After all, it's the good constitution of pigeons that keeps a family together and these have shown a weakness at a far too early age, especially when you have pigeons half again their age full of life, that each year come through a perfect moult, no wonder they were hard to beat when they were on the road.

Fanciers think deterioration comes through hard work. Although I have only flown two pigeons in my whole life that were five years old, during that time they have had to fly hard, and it takes a brave man to take good winners off the road at four years old, but this I have done for the last 45 years. I am sure it is why I always held my own and have never had to rely on old pigeons to keep me at the top. If you are not breeding one or two good ones each year when you have lost those that you have relied on, the loft is doomed.

I never use small seed when birds are sitting the first round of eggs. As I've said many times, the seasons are changing, making the winter drag on to the early spring, and birds will come off their nests for it. Old pigeons know just how long to stay off their eggs before they get chilled. Not so the yearlings who often come off when you open the loft door. That is why I fill the hopper up, plenty of grit and water in the morning, and never go in the loft again till late in the evening. When on their first round of eggs birds need quietness when sitting, especially the yearlings, who can ruin a round of eggs by keeping coming off them every time you open the loft door, but after the first round they settle down and are no more trouble.

Chapter XVII
Observations and Yearlings

Much has been written and can be read on the sport of pigeon racing, but one can never learn more than from observation. Especially is this true of yearlings after mating. All the years I have kept pigeons, my eyes are wide open looking for those yearlings that show keenness and are full of activity. I remember a yearling blue pied; the day after his hen had laid her first egg, he was bubbling over with joy chasing after everything even the sparrows. Other examples were two yearling blue cocks that I noticed were sitting at 7.15 in the morning when I went to open the loft up, although their hens had only laid their second egg two days previously. I am sure there is a good race to be won with any of these three cocks if sent in the same condition.

Among my yearling hens, I had two that, during the winter months I had to keep on parting, as they were always in the corner of the loft together. Although being parted four weeks prior to mating the birds, as soon as I tried to pair them up with two cocks, they went straight together again in one of the nest boxes and would not look at any of the cocks. So I left them together as my memory told me I had had this happen before, and those were the two best hens I flew that year. After they had both laid I took two of the eggs away and when the incubation time came I gave them a day-old youngster to rear and repeated this during the racing season. It is one of the great pleasures when you win with those birds with which you have used common or stock sense, or like the three cocks you spotted prior to racing. These one should not overlook when it comes to pooling.

I know past performance is a guide to most fanciers when it comes to their final selection as regards to pooling for the Classic races. But, only too often the experienced pigeons are beaten by younger pigeons that are being put over the course for the first time. Nine times out of ten, you have noticed the younger birds' keenness before the race, but we all rely too much on past performance and sometimes give back pool money the birds have won in previous years. I am always looking for something else in my loft to beat the good old ones as I have proved over the years a good pigeon when first sent over the distance will pull out its best.

A fancier once wrote asking what you do with a young hen that is barren. I know this is most disturbing especially if they are from your best pigeons and you want a round of youngsters from them. In the past I have had several young hens who have not laid their first pair of eggs in the usual time of eight to ten days. After this time I handle them to see if there is any sign of them laying, seeing if they are high in the rump, low in the vent, with the vent bones slightly open. Failing these signs the hen is given a warm egg late afternoon and nine times out of ten they take to it, even those I am doubtful about. After 14 days I substitute one of the feeder's eggs which has been boiled; if they do lay within the next few days the dummy egg is taken away. There is no need to mark the dummies as they will show up much darker in the nest. In no circumstances mark eggs with a pencil or ball-point pen as this will cause the youngsters to die in the shell. Those that have been sitting 20 days on dummy eggs are given a day-old youngster to rear and I have found that after rearing a youngster for about 18 days the hen has laid without any more trouble.

Using a barren hen

As I have said, barren hens can be most disturbing but sometimes can be a blessing in disguise for their racing ability. I have had three barren hens over the years and each one has more than earned its corn. In fact, the last one a few years ago won a new timing clock in an open race after I had put a pair of warm eggs under her three days previously. This is one of the benefits you get from barren hens, you can have them just how you want them. I mean in the condition they race best at, by putting eggs under them when required or putting a day-old youngster when they have been sitting the required time. I am sure the barren hens must reserve a certain amount of energy by not laying. By adopting my methods with barren or slow laying hens, the cock she is paired to will not take too much out of himself by continually driving, especially if they are yearlings. Cocks at this stage and age are inclined to drive much harder than the older birds, thus using up too much energy. This with the small amount of food they eat during this process can set them back several weeks.

I trust this helps my friend and helps him not to get too downhearted. The amount of setbacks one gets in the sport from time to time must make the novice wonder if it's worth it. But Rome was not built in a day and according to past history the number of times

Observations and Yearlings

BLUE COCK FLASH, NU69L19776.
Winner of seven 1st prizes, 1st Hexham 247 miles, as a YB 5-Bird Specialist Club only two birds on the day, 2nd Stonehaven, beaten by loftmate. Sire the Good Delbar and dam Toey.

the scaffold collapsed when they were rolling the large boulders to the top did not make them give up. When you first come into the sport you get these setbacks and think "why should it happen to me", but these things can happen to the best of us. The longer you keep pigeons the longer you learn to live with the troubles and setbacks you get — like a bad race or being well beaten from a certain race. You act like a bear with a sore head, but before you know it the birds are away again, you time in a good one and you soon forget the previous week. I am afraid the newcomer or novice to the sport is often over-anxious, especially at the vital moment of incubation and rearing period.

Only once in my 50-odd years in the sport did I feel like packing up, that was in 1968 when some scum broke in and stole ten of my birds and they knew which one's to take. It was not so much the value of the pigeons, but to think I had given the thieves hospitality and I had been taken in. It was only through my wife's persuasion that I did carry on. When one reads of fanciers packing up through vandalism or break-ins by the lowest of the low, I know just how they feel!

Chapter XVIII
After a bad race

I have written many times about birds meeting a bad race early in the season, and finishing them for that year. Ever since I have kept pigeons, I have always sent every bird that is fit to our 200-mile race, then I could pick and choose or jump them whenever I liked, with confidence.

With all the years behind me as a pigeon man and I have flown in some very bad races, there was one May race which was the worst I had ever competed in. From my 20 entries I had four on the day, and by the following weekend I was still ten down, among them some of my best pigeons. Then letters started to arrive from Birmingham, Wiltshire and Colchester, all with the same message, "picked up injured". In addition four homed in the same condition, one of my best blue cocks homed with only four tail feathers, cut across the back and chest, with the metal ring embedded in his leg. Yet when he dropped he still had the guts to coo. We sit in the garden and when they don't arrive call them everything or look for a flaw in their breeding!

I was most grateful to the lad of 13 from Brum and the good lady who kept one for two weeks before finding a fancier to report him, also to both the fanciers from Stourbridge and Colchester. Although they would never race again I was always pleased to see them, even if their racing days were finished, as I have found when pigeons are sent back in these conditions the stuffing has been knocked out of them, and they will never race the same again no matter how good they were in the past. They are best put in the stock loft.

Although the seasons are going from bad to worse weatherwise a lot of bad races could be obviated, if convoyers were to get down the line of flight and find out what the weather is, and not rely solely on weather forecasts, which to my mind are only 40% right. One week our two-bird club was at Berwick. The weather forecast was none too good. It was pouring rain in London, but they told me it would spread up north, as the wind was a strong NNE. I felt this rather strange so I phoned down the line to friends from Berwick onwards and found that the rain was only 60 miles outside London. I advised the liberator to let them go at 11.30, estimating a 6¼-hour fly to do the 300 miles. The winning pigeons were clocked at 5.46, and there were eight pigeons nearly on the same minute, but according to the forecast this could have been a holdover.

Bad weather in the middle of the season makes it hard to keep or get pigeons fit for the coming Classic races, especially if one's loft is like mine, well open, and the rain can blow in. Some wet years I had to put plenty of sawdust down to dry out the loft. There is nothing worse than this type of weather for giving the youngsters the snuffles or dirty wattles which, if not taken in hand, will turn to a cold and will soon spread to the old birds. When there is a deep depression over the country youngsters won't run out like they should. I have let my youngsters out on an excellent morning for flying, yet they have come back within half an hour, and I have wondered why, but within an hour or so it has come over bad with heavy rain. This they have sensed and is the reason they have not wandered too far. Get a few mornings like this and it will stop youngsters from putting in two hours' flying and it does not take them long to get into the bad habit of making it only half an hour, which is not enough to keep them fit apart from what they learn when they are away for two hours.

Getting Youngsters Running

One year my youngsters were running for two hours each morning like most other youngsters I saw coming over, then for three mornings running they were back within half an hour, so I kept them in for three days while the bad weather was about and then got them doing two hours again each morning, but had not seen the number of youngsters that were running before the bad weather set in. I am sure this is the reason, confirmed by many of my friends, who say their youngsters have stopped running. Late youngsters that have not started running with the others should be got straight back into the loft, and let out again when the youngsters have come back from running, otherwise they will drop them if the others come back too early.

Youngsters should be encouraged to put in plenty of flying before the first race. When youngsters fly well they eat well, and that's what you want to win young bird races, and put them in good stead for the rest of their lives. Hungry youngsters won't run, if they do and get too far adrift you won't see them again unless they are reported, and nine times out of ten they are no good after having flown themselves out of an empty stomach. My motto always has been to "feed them well and fly them hard".

My second and third birds in the 300-mile Combine were sent back to the same race point on the following Tuesday in the two-bird club. They dropped together taking 3rd & 4th. On the Friday the hen was sent to Selby, 1st NL Fed, that's 750 miles in a week. This was evidently a good yearling that thrives on work, and was one of the two hens that I had paired together. For the record the blue cock 'Alfred the Great' 18739 — that I went to Holland to get, blown over from the Thurso race — the year after was 26th Open LNR Combine Berwick. The next year he dropped two points, 28th Open Berwick (300 miles) 6,684 birds. He had not raced like he did last year, winning seven 1sts in a row before the Berwick Combine. I have only myself to blame putting pigeons like these to 500 miles which takes the edge off them, but he certainly knows his way from Berwick.

While I was writing this chapter, I recalled waiting for my birds to be liberated from our 400-mile old bird race. The birds had been away for five days and I'm afraid my confidence of getting a good one had faded. Long holdovers do not suit pigeons, fit or keen, when sent. They can be beaten by some pigeons which will benefit from the rest and which perhaps get better or more corn than they have been getting at home.

The Prices Paid

I well remember the words of "Tich" Coker who convoyed the Combine pigeons for a number of years telling me how he had seen birds deteriorate in the baskets each day while being held over. Those that are sent sitting 14 days will start to make soft food during the holdover and this will not help them when they are liberated. That's why I try to send my hens sitting eight days or on a youngster the same age, then there is no worry of the soft food going sour. I am sure the longer they are in the basket the keenness and edge you put on them when sent wears off. It's like a pair of shoes that have been left out in the rain, the polish has gone off them. But it helps the prize money to go round.

Look at the amount of eggs that will be found in the basket after the birds are liberated. They come from those that were sent with a youngster about 14 days old. This is a bad condition to send hens for the sole reason of their being held over on the long races. I know some hens put up their best performance at this stage, but one must always bear in mind the possibility of a long holdover and it is most painful for a hen which has laid in the basket to fly 428 miles, especially if the going is hard. I have heard fanciers boast of a hen coming home from a race and going straight into its nest box and laying. I am sure none of the eggs left in the baskets will be from my hens, not under the condition I send them.

When birds are confined in the baskets for a length of time there is a chance of some of them getting "wing lock". These will be mainly those birds that have been jumped from 200 miles; that's why I like to fly my pigeons each week and not rely on jumped pigeons. My birds do not fly around home for ten minutes at most when they are first let out in the morning. I have never believed in making them fly around home and sometimes I get birds come home with wing-lock. If I do I get a bucket of hot water and move the wing up and down making sure the wing is well in the water. This has always done the trick.

Going through some of my old pigeon papers, I came across a 1936 RP which made me think how much prices have gone up: Brand new Benzing printing clocks immediate delivery, £5.50 each; best polished Tasmanian peas 80p cwt. Captain W P Ahern's Stassarts from 'Flying Fox', 'The Duke', 'Squadron Castle,' just to quote a few £7 a pair. I only wish these birds were available today, not because of the price, but because the bloodline of 'Flying Fox' ran through most of my best pigeons in those days. Even so the price was more than the average man earned in a week. My money as a top class glazier in those days was under £4 a week, but the price of pigeons has gone out of proportion.

Take the price paid years ago by my old fishing and shooting pal Jimmy Biss for the 'Scout'. I'm sure the offspring of this could not fly better than his old family which he flew with the LNR Combine. He purchased these for almost a song compared to the price of the 'Scout'. Even that price is nothing today but the price of pigeons today should not worry the novice with a keen eye, or one who is gifted with the full amount of stock sense. There is always a good pigeon to be obtained at a fair price if you know what you want and don't rely too much on pedigrees.

There are some novices you cannot help; those you give a pair of youngsters to and next you hear they have lost them off the top or training. Once a novice asked me if I knew where he could get a good cock. I had some birds sent down from Scotland for an auction sale for me to take to the sale for this friend. I went through the cocks with the novice and found a nice chequer cock that I said I was sure would breed winners. I told him to go to the auction and bid for it. My novice pulled out. I'm sure to this day he let a good one get away for a workingman's price. As I have said, it is not what you pay but what you know as regards stock sense when obtaining your stock or outcross. It's not because it is out of pigeons which have been long since dead.

I can't remember the last time my birds went fielding, but my hens decided to do so after I had wired off all the flower beds where they had been picking about through the early spring. From this I knew they were craving for soil, so I went out to a nearby field, kept a watchful eye out for the farmer, dug half a dozen nice clean turfs and turned one upside down for them to eat the soil. Before long they were waiting for me to give them a fresh one and it stopped them from fielding immediately. Fielding can be a benefit to pigeons if they are eating good clean soil apart from the exercise they take flying to get it.

A friend of mine had birds that used almost to live on a football pitch near his home picking up the worm-casts. He had his best year ever, then. Just after the war I had a caravan at Bembridge on the Isle of Wight, birds from the mainland used to be waiting regularly each morning for the tide to go out to pick among the seaweed. I have seen them pick lumps off and eat them. I bet there was not a fitter team of pigeons. But fielding today can be deadly with the amount of insecticide and fertiliser that is put down on the land. It's not like the old farmyard manure that never harmed anyone; that's why I now always try to stop my birds from fielding.

Chapter XIX
Overcrowding Dangers

I remember the year I was never more pleased than when the old bird racing finished, since a bad race from Northallerton I had been scraping the barrel each week to find a team to hold my own. When you get seven very good pigeons that are finished for the year or will never fly again owing to injury through wires the odds are against you, but it has been said many times I am never more difficult to beat than with only a few pigeons to concentrate on.

One is inclined to get a bit slack when one has a big team and hope for the best. I am sure most fanciers keep far too many pigeons to get the best out of the good ones, taking into account that when you first built your loft it was built to house a certain amount, and before long you have got double the amount it was intended for, and therefore the good ones are not given a chance to show how good they really are. Overcrowding is worse than a damp loft, and has been many a good fancier's downfall. I know of many fanciers who have 200 pigeons and always give you the excuse they have not got the time to race them, if I have not got the time to race my pigeons, I certainly would not keep that amount. Even if one had all the time in the world, even with all my years of success I could not handle that amount. I don't keep pigeons to look at, everyone has to toe the line and give me my bit of pleasure and sport at the weekend.

When I was waiting for my birds to be liberated from Fraserburgh, with a three-day holdover I began to wonder whether their keenness would hold, but I was rewarded with two good pigeons 22nd & 35th Open, 4,448 birds. My second bird was 'Alfred the Great' which I went to Holland to get back. He had always beaten the chequer cock 18724, whenever they met, and I am sure he would have done so again if he had not hit the wires, he was skimmed clean of all the feathers around the keel, yet he was only four minutes behind him. They said I was lucky to have had him reported from Holland, but I am sure I was more fortunate to get him from Fraserburgh, if he had hit the wires across the wing I would never have seen him again. These have shortened the life of many a good pigeon, those that for years have been good winners, and one day you don't see or hear of them again.

Whilst writing each chapter I try not to go over the same ground, if I do you can rest assured that in my mind it's worth mentioning again. After having a long talk with one of my good friends who was most downhearted since building his new loft, having lost 56 youngsters in flyaways in the last three years, we came to the conclusion that his youngsters were not seeing enough of the surroundings outside, like they did when he had a wire aviary for them to fly into, where they could also see the rotation of the sun.

I have written many times how dark my nest boxes are owing to being creosoted each year. This is inclined to make the old bird loft much darker than the young bird side, which is more beneficial to old birds for rearing good youngsters. It also helps them to rest more quietly than if their loft was very light. Not so the youngsters, they need a well-ventilated loft with plenty of light, as I am convinced most flyaways happen when the eye is changing. Whilst youngsters are developing and getting stronger on the wing, the eye change in a semi-dark loft is much slower, and when let out they are very nervous, having, as I have already said, not seen the surroundings and the rotation of the sun. The slightest noise or movement will put them up, and they will be most confused when they are in the air and you will be more than fortunate to get them back.

Although the changing of youngsters' eyes differ in some families according to the depth of colour, light-eyed youngsters change much quicker than the dark eye in your family, also the healthier the youngster the quicker the change. I have never been in favour of a skylight in the roof as this does not reveal the sun's movement and is no benefit to the youngsters and I have never seen one that is completely water-tight. Also if cats can get on top of the loft this will stop the youngsters getting the rest they require at night. Although you would not call my old loft elaborate, it gave me good service for over 35 years and I would not alter it to keep up with the Jones's.

Tea-Cup Critics

One is always open to criticism no matter what sport you are successful at. There are a lot of good fanciers who have never worried about eyesign, and this coming from them is good for the sport in general. But the man who writes criticising one's feeding, training and even eyesign when the only cup they have held is a tea-cup to me has no right to put pen to paper, as it is most confusing to the novices who don't know how their pigeons fly.

I wean off my last round of youngsters the middle of August, and

NO DOUBT
First Open London NR Combine Berwick (8,176 birds). His sire also 1st LNRC Berwick. Great grandson of Champion Mick.

wait for my birds to turn in the dummy eggs before I part them. Then I cannot wait to give the old bird loft a good clean out, especially the nest boxes, and the outside a coat of white emulsion.

Speaking to a friend who won nearly every young bird race in his club the previous year, I asked him how they had flown this year, not one of them had flown the same. I don't know how he flew them but give me the youngster that wins flying to the perch, and not the nest bowl. These you can rest assured will win not only because they are healthy pigeons, but they are born racers.

I know of many fanciers who fly a good young bird, but struggle each year with the old birds, like the novice when he first comes into the sport, everything is in his favour, apart from being dead keen he has a brand new loft free from all vermin, dust and germs which lie dormant through the winter months. He also cannot get home fast enough to train them, making him almost on an even par with the old hand. He gets one or two in the first batch in the early races and begins to think the pigeon game is easy, and the next year they quite expect to do the same, but racing young birds is quite different to racing the old birds, where the old hand has a trick or two up his sleeve. I have seen a lot of novices get very downhearted wondering why their youngsters have not flown the same as yearlings. Although most of my young bird winners win again the next year, those that don't and are just that bit behind after four or five races I stop, and their young bird form has come back as two-year-olds, giving them time to mature and having their first perfect moult. So to the newcomer in the sport don't thrash the life out of them in their second year trying to repeat their young bird form. Take them to 150 miles, and call it a day, and during the winter months you can look round your loft with confidence to some good sport the next year.

Chapter XX
Moulting covers feathers

After flying two years by road transport I became convinced that the young bird moult was much slower than when they went by rail. The heat created by the birds when in the railway wagons was conducive to a quicker moult. Each year my first round of youngsters, which are hatched from my stock birds around the first weekend of March, had, by the end of August when we went by rail, completed the body moult and had two full flights to go with the third one just breaking through and were never in a better condition for the Classic young bird races.

At the same time in those two years not one youngster had completed the body moult and then had three flights to go instead of two. I am not worried about the wing, but I am most concerned about the cover feathers which govern the speed and work they will have to do to get home. With the cover feathers missing, the speed of a pigeon is far less than with them, as the air can penetrate through the wing and the downward thrust is not so effective. The only youngsters I stop when in this state are cocks which are on the big side. As they are stopped for three weeks to complete the moult, I stop them for the season as they have no chance of beating those that have been kept going. Stopping youngsters for one week when heavy in the body moult will make no difference as by the next week the condition has not changed enough, but the distance has.

I like to keep my youngsters going as the programme is too short even to miss one race. The chances of winning with a pigeon in this state of the moult are far less than with one with full cover but they will get home. True, when they meet a belt of rain they are the first ones to get wet owing to the rain penetrating. They will drop and wait till they dry out and come on again, but all the time they are being educated, which is far better than being left at home. I am speaking of the shorter races, around a 100 miles and would never send them to the final Classics in this state. You would not only be throwing money away but also youngsters. The only time I go through my youngsters is a week before our Combine race and only those that I think will do me justice and are nicely cleaned up do I send. I have won with youngsters in all states of the wing and tail, but never with the cover feathers missing.

After a hard fly youngsters are best kept in the following day. Even those that come late on an easy fly are caught and put in another loft while I let the others out as they would only bring the team back from running. They will stop the birds having the amount of flying they need not only to keep fit but to keep them at the right weight to win. Stop a youngster for one week and see the amount of surplus fat it has put on, and when put back on the road with the cover feathers still not fully grown they make hard work of it, even in the shorter races, and can run into trouble. I have never been an envious man in regard to pigeons. I have had all the sport and pleasure one could wish for and when you've done it all by yourself all your life the game gets harder.

Up and Coming Fanciers

There are up-and-coming fanciers but they need the help and encouragement from the older ones, as there are not enough of this type coming into the sport today. There are many other activities for them, which can tend to put them on the wrong tack, but my mother used to say she always knew where to find me — among pigeons. I spent my last 10p at the age of 14 to enter two birds in a Perth race and had the only bird home on the day in the Sunday morning club, where we all timed on the same clock, with your name on a piece of blank paper already in the thimble. All other birds were recorded on a card when the clock was filled.

This brings me to one of the most memorable occasions of my early days. A fancier came tearing out of his house with the thimble, running to where the clock was positioned. At the same time a copper was walking down the road. Thinking the fancier was up to no good the policemen gave chase and brought him down with a rugby tackle! There they were, looking for a thimble when the fancier told the policeman what it was all about. Nearly every house near where I lived as a boy kept pigeons and the sky used to be dark with pigeons when they were let out. Birds used to change hands for a few pence and more often than not within a few weeks were flying to a new loft. Not all would break in that easy and after several attempts the fancier would want his money back, but there was always a customer for them.

Experience has since taught us that the brainier the pigeon the easier to settle. I have had birds that I have parted with come back once and, on finding its mate was not there, have left the next morning and found their way back to their new loft sometimes flying 200 or 300 miles, sometimes on a route they have never been trained on. Broken pigeons that have flown out all the summer with their mates as soon as parted will sometimes return to their old loft and as most fanciers who part their pigeons do so during the autumn months, the reason why some never make it is because of the amount of mist and fog about at this time of year. This will beat the best of them.

More two- and three-bird specialist clubs are springing up all over the country some flying midweek. This is the best education you can give your pigeons. They learn more in one of these races than with the Fed on Saturday each week. With only 30 or 40 pigeons they can see more of the surroundings and nine times out of hen have to rely on their own brain and not on the bulk of the Fed birds. One is always trying to make them individualists. Flying midweek with a handful of birds will eventually achieve this and when the time comes they have broken well from the race point you will reap the benefit as they have gained confidence in using their own head. There is no keener competition than that provided by specialist clubs. Members only send their best when they are limited to two or three birds. No club can do without the mob flier — he makes the club money, but asking fanciers to pick out two birds really determines how good a fancier is.

I remember receiving a most disturbing letter from my Dutch friend Mr Nebbeling of Dan Helda, who told me he had in his possession over 40 English pigeons, that he had reported to the RPRA and the fanciers did not want them back. Why should someone else have to feed or dispose of other fanciers' rubbish? I say rubbish because real fanciers would be most grateful to have them back. I'm sure the birds friend Nebbeling got in are those that I have heard said many times have been given their last chance in the Scottish races, having made many mistakes before, as it is from these races that birds sometimes get blown across to the Continent. When the entry fee becomes too much for them to send their rubbish, men like Mr Nebbeling and the RPRA will be less troubled by the stray problem and we will get better races.

Chapter XXI
Bad Trapping

Among the many letters I receive from fanciers, one in particular stated that his birds have flown much better on hopper feeding, but went on to say that he would have done much better if his birds had trapped on arrival. To the first part of his letter I am pleased to think that he has tried someone else's methods to improve his birds, to the latter I am afraid I cannot agree with him, as nine times out of ten bad trapping of youngsters is due to bad management or not educating them when young. This I have covered most carefully in a previous chapter, and can assure you that to avoid it I spend more time at this stage and age with my birds than at any time during the year.

There are different types of bad trappers. There is the one that on arrival goes straight on the house. These I have no time for as I am sure they have got in the first batch by luck, and not by the way they have left the race point. No one can convince me that when a youngster breaks well from the race pannier with every intention of racing its heart out to get home and having been well schooled it chooses to stay out. True there are some no matter how well they have trapped before racing starts, as soon as they have that night in the basket and are liberated with several thousand birds, get keyed up, upsetting the nervous system which is a characteristic in the family. Knowing this, one must bear with them with all the patience of a good fancier and let them come in in their own time, all the flirting with drop birds won't make a bit of difference, in fact, it will only make matters worse.

Then there are days while waiting for your birds you can almost hear a pin drop, not a sparrow or a wild bird to be seen, everything is still, like the lull before a storm, even with the old birds one can expect trouble at the home end with the amount of electricity and stormy atmospherics in the air making the wild birds seek cover till it passes. But I'm afraid a lot of pigeons are made bad trappers by the way they are caught on arrival. Birds that go straight to the water should be allowed to drink before catching. Although you may lose a few seconds, you will gain more in the long run.

Never try to catch youngsters in an empty loft. There are always some youngsters which are not quite right when you basket up at the weekend. Leave these in the section, failing this a few old hens a bit on the hungry side put in before they arrive will steady them down when they come in. Don't try trapping youngsters or even old birds in something they have not seen you in before, I am speaking from experience which cost me the London Championship Club. While awaiting my birds I asked my wife to get me a pullover as the sun had gone in and it was a bit on the cool side. Not thinking, I put it on, and when I tried to trap my first arrival it shot off the loft like a rocket. Knowing what was wrong I took the pullover off and trapped my second bird with no trouble to be 4th Open, but my first arrival never forgot, he stayed out for three days. Each time he came over the loft he would cock his head on one side and look straight at me, I called him the 'Nervous Cock' yet he was bred from my most tame pigeons. I am sure that is why when we visited the maestro of the Belgians, George Goossens, we were given a brown coat to wear, the same as the one he always wears when pottering about in his loft. The last time I visited him I had four friends but he found us all a coat.

Feeding Acorns

All the years I have been giving my pigeons acorns I have never seen them as big as they were one year, which reminded me of several inquiries, how do I feed them to my birds. It is obvious they must be crushed up either through an old mincer or between two bricks and put in the trap or aviary or even in their hopper. The old birds need no enticement to eat them as they know what they are but the youngsters are not so sure but it does not take them long to get the flavour, as one can tell by their droppings the next morning, being of a darkish nature full of oil. Make sure the acorns are not wet when you collect them as they will go mildewed, they must be used up straightaway as they will not keep in the sacks unless you take them out and put them where the air can get all round them.

Do youngsters that have had only three or four races make better old birds than those that have flown the programme? That is a question I am asked most frequently. Although I do not stop youngsters half-way through the season hoping they will make better old birds, those I stop are for stock purposes according to the eye and type. I am convinced that those that fly the race programme will hold their own against those that have only had two or three races. Although I am not in favour of sending all youngsters 300 miles, I have had some good 300-mile winning youngsters that have made good pigeons, but they have been fortunate to have a good day, being on the wing eight hours, which in my mind is quite long enough. The same applies to making them fly the whole programme, it all depends on the type of season they are having and each year fanciers think it is harder than the year before. Seasons or races only seem harder with the type of youngsters you have bred.

If you have bred a bad lot and each week drop one or two the season will seem harder, but I am only interested in those that I have left after flying the whole programme when the season ends. I can honestly say I don't lose many youngsters racing but don't think I never breed duds, on the contrary I breed my share, and these I put down long before they reach the race pannier so they can't drop by the wayside when things get tough. I am convinced that 75% bred each year are duds as racers, 20% are fair pigeons, leaving only 5% the very good ones. What makes a good pigeon stand out among all its loftmates in one season? Is it because it is your first bird each week? Is it because it is a perfect pigeon, well balanced, more brains than its loftmates? Has it inherited more of the winning genes from its parents or grandparents? Is it physically or mentally fitter or has it found a good line either as a young bird or a yearling? Obviously all these qualities help to make the champion, but I am convinced that the last is more the answer, they have found a good line and stick to each week. When youngsters get a bad line nothing will alter it for that year.

Getting a Good Line

My first youngsters in the early races are always my best right through to the Combine and one can only hope that the brain or homing ability has developed more when next they go on the road as yearlings, which will enable them to find the right lines. Most of my best pigeons have come from west of the race point. On the North Road even with a strong west wind they have come the same way to win, they may lose a bit of ground flying into it to get on their line but will soon make up for lost time, and only these can we call individualists, sometimes flying miles from the race point on their own, when the bulk of the birds come from the other direction. But they are few and far between, and come only from the 5% of birds bred.

When the season is over a lot of fanciers will be wondering why they have not done any good and will straightaway condemn the pigeons, but first look at your loft and self for the trouble. Is the loft well ventilated, free from draughts, weatherproof top and back? Have the birds been happy while in the loft? Only to mention a few of the many things that can be your downfall, and above all can you breed a good youngster? No matter what pedigree your birds have, if you have not got the knack for breeding good healthy youngsters they will not score. I don't know of any fancier who bred a better youngster than my old friend the late Frank Palmer, it was always a pleasure to visit him when he had weaned them off, they were a joy to look at. Although he had no special preparations the parents had ample food in front of them consisting of all different grains, plus the regular cabbage which they were continually picking at, and no one can say he was not a hard man to beat. He loved pigeons inasmuch as he would not put down the weaklings. He would say: "It's not their fault they are bred like that", and I don't think he ever killed one, but he would have been a much better fancier if he had suppressed the weaklings. In fact, he had a medicine chest full of remedies and cures, only because he was soft-hearted. He had a saying "No happier kind you'll ever find". It was true not only among his pigeons, but to the novices who owe their success to his kindness and generosity.

They pulled down a disused railway station at the back of my old loft which must have been infested with mice, and they started taking refuge under my loft. I noticed that the core of the maize was eaten from the hoppers I left filled at night. There is nothing worse than mice wetting over your corn and disturbing your birds, but with the modern Warfarin, it is easy to get rid of them.

Chapter XXIII
Extras in the loft

I was sitting having my lunch when I saw a youngster fly up onto the aviary which is about three feet high so I ran down to see which one it is and make a note of its number as nine times out of ten it's a winner. The cheapest thing we give our pigeons is grit and I have seen this in lofts for days, I give them fresh grit everyday and throw the rest away. Green stuff is most essential, my number one is dandelions, number two watercress then a cabbage. I chop up a bowlful of dandelions every day root and all. I find they like this best of all, I put it on the lawn to save it getting dirty inside the loft. I give this to my young birds in pots when I have cleaned the loft out in the morning. When I go to work on a building site I always have a clean small sack with me to collect virgin soil, from mole hills or fresh rabbit holes or where they dig deep to put in posts or fences. I tip it on the path after I have swept it, and I have come home after three or four hours and found it has all gone. Pigeons love clay-soil; why I don't know it must be the minerals in it. During the war we had like most people an Anderson air raid shelter at the bottom of the garden and the top was packed with clay and soil. They used to spend hours on it pecking away, they ate right through down to the metal and the youngsters droppings were nearly all clay and they were some of my best youngsters I have bred and the birds flew well.

I remember one time they were over the back of the shelter where the weeds were growing so I crept round to see what they were doing and among the weeds were some deadly nightshade. The ones with the black berries. They were picking them off like hot cakes I nearly had heart failure I thought they would all die. That week I took 1st & 2nd LNR Fed so it did not hurt them. But I took no more chances and pulled them up.

A novice whom I had got very friendly with said he had spent a lot of money on pigeons and was wondering why he could not do any good with them and asked me to have a look at them. When I saw the pigeons and handled a few my words were: "They are a lot of rubbish and even I could not win with them". I can see his face now, it dropped and it was a few minutes before he could speak. I said I was sorry if I had hurt him but that's what he asked me over for and he said he would get rid of the lot. Then he said I have just brought the last ten pigeons off a fancier who had packed up, and would I have a look at them. When I saw them straight away I said these are better and was keen to get my hands on them. When I went through them I said "Pair that blue cock to that red hen and they will breed winners in every nest". They bred fifteen 1st prize-winners including 1st Open Western Home Counties Combine winner from Berwick 308 miles and 2nd Open Combine Thurso 507 miles. His loft was founded on that one pair of pigeons.

Eyesign in Books

I don't want you to think I am being boastful but I am trying to prove to the fanciers that poo poo it, there is something in eyesign, and as I have often said if they can prove to me there is nothing in it I am willing to listen but they cannot — they do not know what they are looking for. As I have said I have never read a book on the subject and have learnt from my success. We can all learn from our mistakes but you learn more from success and as I have said you cannot put it on paper. I once remember having an eyesign book sent to me from South Africa and in the first chapter it said pigeons with egg-shape pupil are degenerating and should not be bred from. My good blue cock 1140 won 1st LNR Fed, velocity 795 ypm, the next week 3rd Open LNR Combine as a young bird, a daughter from him previously mentioned 'Silver Blue' hen bred four cocks to win over £2,000. His pupil was egg-shaped from 2 o'clock to 8 o'clock so it was thrown straight on the fire. That is the only time I have looked at a book on the subject.

One Cock; Five Young Hens

Getting back to the young birds, the performances of my young birds were second to none in London each year. Not all families win as young birds. Most times they are kept too hungry but mine have all the corn they require and were always ready to be played about with. I have had five young hens sweeping up to one old cock in the old bird loft on his own when all the other birds were shut outside for about two hours. This was in 1958 before the LNR Combine Young Bird race.

As all the nest boxes were empty he was taking them in and out of each box till they were fighting over him. Then I basketed them up and took them for the race the next day. They were up about 9.15 and I reckoned about 2.15 I wanted a pigeon. I told my friend the time

THE LAIRD, NU67L12685.
Winner of twelve 1st prizes, twice 1st Fed, 96th LNR Combine Open. Grandson of Champion Mick.

they were up and he reckoned the same time, as he used to get them in with a stick when I went into the loft so he said he would be round about 2 o'clock. Just before 2 o'clock a pigeon was coming straight off the road. It was one of the young hens it dropped like a stone. In the loft I went but it hesitated and I heard my wife say: "Tony go and help your father with that pigeon, it's not going in". So he came down the garden and I said pick up the cane to drive her in. She was clocked 2.01. I came out of the loft and looked out to where she had come from and four pigeons were coming straight for the loft.

They also dropped the same way. They were the other four hens clocked 2.03. They were 1st, 3rd, 4th, 6th & 7th Open LNR Combine, 4,631 birds and I had all my money on them. My friend arrived just as I was coming out of the loft and he knew by the look on my face I had timed in. The winner was my 'Blue Girl' daughter of 'Champion Mick', seventeen 1sts, four times 1st LNR Fed London Championship Berwick, 4th Open LNR Combine Berwick 4,668 birds. I would like to point out the cock must be getting on in years also he must be one that is not spiteful, one that just gets down and calls and lets the hens sweep up to him. A good thing to do is to let three or four old cocks out with the young birds and see what cock the hens all want. There is something about some old cocks some hens like and that's the one to use. Young cocks can be given that little bit of extra to race home.

Rank Old Hens

I always have a few old hens that are rank, I let all my old birds out and shut down the front so they cannot get back in. I bring round the old hens and put them in the old bird loft then I put as many young cocks in with them before basketing about two hours. These must be those that have been most consistent during the young bird racing. Watch the ones that are in the nest box with the hen and put your money on him. Both these systems I only do in Open or Combine events or if there is something worth winning.

I don't encourage my young birds to lay but if I find a young hen on the verge of laying I give her a nest bowl and a week or so after laying they are hard to beat. If you don't put the nest bowl there the egg will be laid on the floor and smashed. I have often been asked if young hens are any good after they have laid – this I am sure depends on the condition when laid with regards to moult. When young hens have laid after they have cleanly body-moulted I have won with them as yearlings, but have not had much success with those that have laid before they have body-moulted.

Of course I have had plenty of success with cocks and hens sitting six to eight days. My good mealy hen won over £600 sitting, my good blue hen 'Scottish Lass' won fourteen 1sts, 1st Open LNR Combine Fraserburgh sitting same time. I like to have some sitting and on the jealous or semi-Widowhood system for the Combine races. The reason the Widowhood pigeons do not do so well from the long races is they are in the baskets too long before liberation. If they were flown to the racepoint the night before the long races and liberated in the morning they would be hard to beat, or the day after, but after that it has worn off and they are not so keen as those that are sitting. Plus they are on the light side when sent owing to not eating as much as the others, and they also lose weight in the basket. For the longer races birds, in my opinion, should be on the fat side allowing for them to lose weight in the basket. I don't like big pigeons, my cocks are on the medium size and my hens are on the small size. I remember marking for the Thurso race one of the smallest pigeons I had handled, the only one this fancier had sent, it was pooled right through and I said to myself if that's a hard race he won't see it. It was one of the hardest races the LNR Combine had flown, liberated 5.15 only two birds home on the day and this one timed in 9.55 at night for 2nd Open Combine.

Never again did I condemn a small pigeon. I have since found out that medium and small pigeons don't need so much work to get fit. I have never forced my birds to fly, they have an open loft and are always taking off, sometimes they will go running for over an hour. I think training of old birds is over-estimated and I am sure you can make them sick of the basket. If they are trained well as young birds plus about six or seven good tosses from about 35-miles before racing after that they are best left alone during the week, unless they have not had a race for a week or two, the couple of 35 milers in the week will shake off the cobwebs and bring them right again. I think yearlings want a little more training than the older birds. My good mealy hen WG78 had four tosses at 35 miles and staight into Berwick Combine 300 miles, 1st Sect, 7th Open, she had already flown out of Scotland ten times up to four years old. I had intended retiring her at her age as she had already won over £500.

Experienced birds and big jumps

The point I am trying to stress is that old pigeons which have flown the distance can have big jumps. She and 'Scottish Lass' were the only two pigeons I flew over four years of age. I think when the fanciers have to rely on old birds they are not producing good birds every year. If we can only breed one very good one each year we are lucky. All my best pigeons I have stopped at five years old and each year I have had something to take their place. Every year you hear of someone saying they have lost their old so-and-so. You will lose your best if you keep sending them and nothing is more distressing than to lose an old favourite. I say when they have done all you have asked of them stop them and build your loft around them and don't be too greedy or you will wish you had not sent them that once too often.

When I read of the great grandsons of champions, that have long since been dead, for sale, the chance of reproducing them is nil. I think grandchildren are as far back as you want to go to reproduce a certain pigeon. After that so many other genetic factors can be produced, I am sure a lot of good producers have been killed because they are useless on the road.

Take the Silver Blue Hen 1721, she was one of my best stock hens. Her sister won over £500 but never bred a winner, but the Silver Blue Hen had the breeding eye, and from this eye I had bred winners before. It is a violet eye with a grey circle, that is why I kept her. Also she had been picked up and had therefore not gone in the first loft she saw when distressed. I don't like birds that have gone in another loft. Most of the birds I have reported are picked up in the street. I cannot remember the last pigeon I had reported that had gone in someone else's loft; this kind I don't want. I remember sending a good blue pied hen to Berwick 300 miles as a yearling, the race was a tricky one, returns very gappy, she returned the next morning with a note in her ring. Come here I said to myself, you have been in, but when I read the note it said picked up and the fancier who picked it up was a member of my club and only lived a mile away, so straightaway I went to where I knew I would find him. I said what do you mean by putting notes in my pigeon's ring. Straightaway he said, "Was she yours", and he said whilst he was on his allotment she had dropped quite close to him at 9.45pm so he picked her up and took her home with him. He said she looked so well the next morning he knew she would get home as she was one of the local club's pigeons.

I sent her to Newark the next Saturday which she won easily plus 2nd Fed. The next Tuesday I sent her to Perth 350 miles for the race the next day, they were up at 7.15am she dropped with my other three entries at 6.20pm and was trembling when she dropped, the fittest one of the four, they were just beaten but I was pleased to see the blue pied hen with them. She had made amends for Berwick, I shall never know but I am sure she had gone over the Continent as the wind was strong westerly and she started making her way back home into the wind and she just could not make the extra mile. I am sure there is a certain amount of telepathy between the good fancier and his pigeons.

Wind and Strange Ground

The wind is the main factor in racing and I have never bred one to beat it. If there is a strong west or north west wind the fanciers on the east side have a big advantage, and the winners are always on that side and vice versa. I think the true race is a head wind. In 1968 I won the Five-Bird Specialist club Gold Cup Best Average all races, four were a strong north east wind, being the most westerly fancier this was to my advantage. Also they would come down the line I had trained them again to my advantage as the others were on strange ground. I won all four, each time a batch of 12 to 14 birds came straight over my loft out of the west. It was obvious that they had been brought down by my birds as mine were about 40 yards in front of the batch.

I am often asked how pigeons home, it has always been my opinion they home on a wave beam. I like to think of a pigeon's brain as a radio receiver, the pigeons that win up to 300 miles have a three-valve receiver. The pigeons that win over that distance a five-valve receiver, they pick up the loft signal much quicker. I am more sure my theory could be right as each week pigeons are not homing like they did years ago with the amount of tall buildings that are going up all around London and the big jet aircraft that are in the sky today breaking up the wave beams I think they home on. Take the reception from the television set, when interrupted by a tall building the reception is bad and this is what I think happens to the pigeon brain.

But there is more to the long distance pigeon, he must have plenty of stamina and a good heart and not go too hard at it straight away, that's where the good, keen 300-mile racer goes wrong. The good 300-mile racer goes all out straight away and runs out of steam unless it is a fast race but when we send to the long races for a 12 to 14 hour fly, the one that takes it steady will be hard to beat. But we must have the keen 300-mile pigeon, I think that's where a lot of the long distance men fail. They introduce another long distance pigeon and they get plodders. No you want to pair the keen 300-miler to your best long distance birds. You must not lose that keenness and urge to get home. I think we can call them races up to 400 miles; the others I call endurance tests or obstacle races. Look at the fanciers that win them, and the pigeons have done nothing for three or four years then up they come. The good fancier is not tarnished by being beaten by a pigeon that has never won before in its life. The further you go the more change there is in the weather and leading pigeons come up to a belt of rain and go round it, those that are half an hour behind on the same line when they get there it has stopped raining and the leaders fly miles off their course.

Good and Bad Breeding Year

I think 1949 was the best year for me, I bred seven champion pigeons that won seventy-two 1st prizes, twenty-one Combine positions including 2nd and 6th Open Combine Fraserburgh, previously mentioned, only 18 home on the day. In 1950 I bred two very good pigeons one was 'Red Admiral'. In 1951 not one, 1952 the same. In 1954 two very good pigeons including 1st LNR Combine with 'Denque'. Again I am not being boastful but trying to quote what I mean and this often happens as the winning genes comes in cycles. So many things can happen when breeding and rearing bad or wet feeders, corn a little musky or mice tainted, but the main thing we all hope for is the right genes when the cock treads the hen. I once paired two pigeons together and made them my number one pair, I could not wait to see what they would breed. When the youngsters were weaned I was disgusted with those two horrible little smokey blue hens, the next nest the same, they were throwing back too far for my recollection so they went down again and I forgot all about them and when I rung them I knew they were different and when they had been weaned off they were what I expected, two lovely blue cocks. They were together in the young bird loft and I knew the true parentage but I have often found that late breds from a pair that have been together

THE GOOD MEALY HEN
Winner of twelve 1st prizes over 300 miles, 15th Open Berwick London NR Combine (6,163 birds); 7th Open LNRC Berwick (6,761 birds); 50th Open LNRC Thurso (2,527 birds); 18th LNRC Fraserburgh (4,428 birds); 55th Open LNRC Thurso (3,877 miles). Daughter of Red Admiral.

all the year are more true to type. This I don't know why perhaps it is because they are more adapted to one another or it's third time lucky.

To breed a champion from a pair of pigeons is like a roulette wheel, 49 blacks and one red and you are trying for the red each time which is the winner. I don't breed from pigeons that have won once or twice, I like them to win at least five times, so I know they are genuine. When you get one that has done just that run him to all your best hens, not too near related, putting down feeders at the same time. I founded my loft on the 'Plum' mealy cock, I had him paired to three hens in the young bird loft at one time. After five days together I let them out with him for four hours each day and they laid within four days of one another and four of them were 1st prize-winners. Good producers are wasted rearing youngsters, I only use five or six cocks to breed from each year pairing them to my best hens using the others as feeders.

Chapter XXIII
No close inbreeding

I should like to think I have bred and raced as many good pigeons as the best, but never once have I tried to reproduce any one of them, in fact I'm sure I wouldn't know how. By that I mean close in-breeding, to quote: "Mating brother and sister together, pairing this back to that of the same family". The outcome or the percentage of winners would be minute compared with those of the first cross, which is quite the opposite to close in-breeding. I would advise my novice friends not to indulge in it, as the outcome could be most disappointing. I am sure you cannot call uncle to niece or vice versa close in-breeding, which has given me my best results in producing winners, when trying to reproduce the pigeon in question.

Too many fanciers rely on the outcome of the youngsters from the pigeons that have done the work. Among the good pigeons I have bred, not many have been good producers themselves. I have had my best results from daughters of my best racing cocks and from the sons of my best racing hens. When looking for a cross I would rather have a son or a daughter from the pigeon that has done the work than purchase the original article.

I seldom breed from my yearlings until they have shown some characteristic sign of the parents, especially in likeness. I have often mistaken a son for his sire, or a daughter for her dam. These I have found have produced the winners almost identical in shape and type. When it takes two or three years for a pigeon to develop one is taking a chance breeding from yearlings. I can look at photos of my winners taken as yearlings and there is no comparison with the photos taken two years later, one would not think they were the same pigeons, developing a bit more size, wider across the back with just that extra length making them perfectly balanced, with a head and tail that fit the body, especially the latter.

I am sure fanciers do not pay enough attention as regards the tail, and in my opinon, it is one of the main factors in the make-up of a good pigeon. The tail should belong to the pigeon as if it had come from the same mould. The quill should be almost as thick as the primaries, and when in the hand the tail should drop over the finger next to the thumb with a gripping feeling, and should look like one feather, not with the fan open and definitely not look as though it has been stuck on after the body has been moulded. Only when all these things materialise do you know what you are breeding from, or does the resemblance of the champion occur, then you are on safe ground for reproducing.

No Colour Prejudice

I received a letter from a fancier asking me if I liked grizzles. Many years ago I had a good Bricoux grizzle hen, not only was she a good winner but I bred three blue grizzle cocks from her that won 1st prizes in turn and that blood is in my loft today. I remember the words of the late Bill Vaux who lived in the next road. His loft was typical of the Continental style, over 14 feet off the ground, saying he had never seen that colour do much good on the road. Coming from such a good fancier as him made me wonder, but it was not long before he had to eat his words, when the three cocks dropped together to take first three in the Fed. I remember him coming up the garden ten minutes after timing in, saying: "Who said those bloody grizzles won't win?"

I have never been prejudiced against colour, as long as heart and brains are in the right place, the colour won't stop them from winning, but there are colours and colours, the chequering on blue chequers should stand out most prominently and not as though they had been smudged when wet after being painted. Blues should be of a deep colour and not as though they had been through a washing machine several times making them almost powder blue. The same applies to reds. The chequering should be like that of the blue chequers and not wishy-washy. I have heard fanciers say they would not have a mealy in their loft for anything, but again, there are mealies and mealies. The best mealies I have had are what we in London call a strawberry mealy with a depth of colour, not washed out, almost white. I have in one of my previous chapters dealt with whites, there are good and bad among them, but depth of colour among them all is the point I am trying to stress.

Fanciers ask me where have my reds and mealies gone? Without the reds you lose the mealies, and I have only had one or two red pigeons at any one time, in turn they have invariably bred mostly blues and blue chequers, the colour base of my original family, so I stuck to them. My good red chequer 'Red Admiral' only bred four red pigeons in 14 years at stock, mostly blues and blue chequers, white flights with the occasional mealy again with the depth of colour. His sire was one of the best violet-eye cocks I have ever seen, not only in my loft but elsewhere. He was my 'Blue Lad', fourteen 1st prizes, four times 1st Fed; 2nd Open London NR Combine, Fraserburgh; 6th Open LNRC Thurso; sire of twenty-one 1st prize-winners, a son of the 'Plum' cock, as the dam of the 'Red Admiral' was the unrung hen I had on loan. I was most pleased each time he bred blues and blue chequers, white flights, showing that my own colour and strain was most dominant.

I am not trying to dazzle readers with my performances, but trying to emphasise the fact that I like to produce the same colour as my best pigeons. Many years ago I was talking on the subject of eyesign, when a fancier showed me the droppings of one of his pigeons that he had taken from the basket saying, "If this is not right — all the eyesign in the world would not help them to win". He never spoke a truer word. Much can be learned from this as regards health and fitness. I often wonder how long it takes those who don't scrape out each morning or those that are on deep litter, to find out that one of them is not right, and stamp out the trouble before it has spread, not only to their own birds, but those that come in contact with them during the racing season. Only the other day a fancier told me that his birds were immune to diseases, being on deep litter over a number of years. I take that with a pinch of salt, that does not mean to say they could not be carriers. I'm not saying the scraper is a way to success, I know some fanciers who keep their loft in a deplorable state and win.

No Fear of Hard Work

It has given me much pleasure to air my views in this book and also the opportunity of showing some of my great pigeons of the past and hopefuls of the future. Sixty years ago my life's ambition was to fly a good pigeon. When I look back now all was not in vain. True, I have been gifted with a certain amount of stock sense, but I also had the essential thing in life — the will to win. To those who do not agree with all I have written, I know I may not always be right. There has been a great deal of criticism of the deep litter system; I have never been against it, as I know only too well the number of fanciers whose pigeons fly well under these conditions, but I like to clean out each day the same way I did when I had my first success.

I have never been afraid of the hard work that is involved when keeping pigeons and only from this have I gained my success. I have written many times of the hazards and enemies the pigeon encounters en route during racing and home, one in particular is lurking in the loft day and night, and that is dust which cannot only ruin a pigeon's health but the fancier's also. Every pigeon has a certain amount of saliva at the back of the throat which is there by nature to stop the dust penetrating the lungs, in a very dusty loft the saliva builds up to give resistance against it, and in doing so will stop them breathing naturally when on the road; they will soon run out of steam and will have to give up.

I have said many times, excess use of the scraper is not the way to success, those like me who scrape out their perches once a day and the floor at weekends, sometimes forget to brush the perches well and sweep the floor, which in my mind is most essential. Before doing so I damp down some coarse sawdust, adding a little Dettol to the water, then sprinkle it over the floor making sure I clean out all the nooks and corners. This not only stops the dust from blowing back on your perches, but makes the loft smell sweeter, and when fanciers are talking about the well ventilated loft this is the answer, letting the dust escape outside.

Have we ever really stopped to think how some fanciers have to keep their pigeons under difficult circumstances? By this I mean the ordinary back-yard fancier, I have seen some back gardens in which there is barely room to erect a loft, then there are those with neighbour troubles, plus the cats — these are only a few of the many problems they have. Yet from these come some of our best fanciers. I often wonder what their performances would be if they had the best facilities to keep their pigeons. Talking to a senior partner of a partnership who admitted that his partner's time with the pigeons was very limited, yet the partner spots more than he does as regards fitness and the many details that go to make a good fancier. This does not only apply to partnerships but to many who read all they can on the sport, yet when it comes to putting it into practice they fail.

I am convinced that if a panel of top fanciers were to talk all night on their ways of success to an audience that would fill the Albert Hall, those who had absorbed the main items would be very few; those that when it comes to putting it to practice would be far less. The same applies to different trades, some men pick it up much quicker than others and some never do.

Training in tricky weather

Turning to practical things, in the autumn, the good fanciers will have washed and repaired their training baskets also the sacking which can get caught up on the pigeon's leg and cause it to come home lame. I am afraid a lot of fanciers overuse their training baskets to find success. Little do they know you can make pigeons sick of it, especially if you stick to the same line of flight, that's why I like to train in all directions — not only does it help them to get home when they are driven or blown off their line, but it becomes more interesting to them. Most fanciers when they send their pigeons training, pick the perfect day from which the pigeons have learnt nothing except they have stretched their wings. I have trained pigeons some days when most fanciers would not let out. I like tricky days when birds have to use their heads, good visibility with broken cloud and rain or heavy showers about hoping they run into one forcing them off their line or with low cloud with fair visibility and the temperature normal meaning not cold, and put up in twos or threes, they will learn more from one of these than a dozen of the fine-day tosses.

Pigeons' brains soon become dormant if things get too easy, and as soon as they run into trouble it takes them a day or two to get home. It sometimes takes a bad toss or race to bring some or most pigeons' brains alert, thus brushing away the cobwebs that form when things have been too easy. How often do we get pigeons that have made a mistake on training or a race coming back a day or two later topping the Fed the next week? One year I sent my four South Road entries for the YB National to a toss 40 miles north. I had three together about the right time, the other one came next evening. On the Saturday that one beat my other three entries by half an hour, I was 164th Open, proving that the flying he did and the ground he covered put the homing instinct right.

Ever since I started keeping pigeons I have always looked round to improve my ways and birds and still do, either by a good cross or by trying different feeds during racing to get the best out of my pigeons. I remember making some malt pellets, the next morning they had all run together like a slab of toffee. I even had my own ingredients for a speed-cake, which was to be kept most secretly even from my children, which consisted of a dozen things including two new laid eggs, and an eggcup full of sherry. The birds flew well on it but the next year they flew just as well without it, but there is nothing to beat good clean corn, a happy loft of pigeons to win races. There is never any harm in trying out certain ideas one hears or thinks of. On the contrary sometimes a lot of good can come from them, but above all it shows keenness. This alone can help you to win but over-keenness can do more harm than good. One in particular is the breeding of early youngsters, I know several fanciers who have youngsters weaned off before I even pair up. There is nothing to gain from breeding early youngsters, only making the hard work sooner, and the early bred youngsters stop running long before the main event where the money is. While those bred around late March or early April will still be running much nearer where it counts and will be much fitter in mind and body. Apart from the main thing they are not too far gone in the wing moult.

Chapter XXIV
Winning with Plum

I think successful pigeon fanciers are born, with a gift for the knowledge in them. I used to spend hours watching every movement and their ways. You will learn more that way, than reading about them. I never see if a pigeon is fit for a race by handling, as I've found this will let you down. Some of my best birds never handle hard and firm when ready for a race. My champion 'Plum' mealy cock never carried any flesh, but he won seventeen 1sts, eight of them in succession.

I shall never forget the last race he won out of the eight from Berwick (300 miles). The late Wally Dimmock said, "Sending 'Plum' again? You'll lose him," to which I replied, "I bet I don't. Never stop a winning bird." It was only a figure of speech, but Wally being a betting man put money on the table. "Cover that," he said and before I could say a word, they all wanted to have a bet. Well, I took a lot on 'Plum.'

The Saturday morning opened up fine, birds were up 7.30, wind SW. I knew it was going to be a hard fly. At 10am it came over black — started to rain and never left off all day. I was sitting down having tea when my daughter said, " 'Plum', dad." There he was, shaking the rain off. I clocked him in 5.32, 1st Club, 2nd Fed; what a relief! I made a fuss of him, not because he had won the bets, but because I could still hold my head high, and he proved what a great pigeon he was.

You may ask how he kept on winning. I had him paired to his mother; she went barren the year I bred 'Plum' and he was always after her, so I paired him to her and sent him calling to nest first race. When I came back from the club I put a pair of eggs under her. He used to come home and go straight on the eggs till I took them away on Wednesday morning to get him calling to nest again. He won the LNR Fed five times and all his races were won in this condition. I find that if a cock races that way, so will his sons and grandsons. My 2nd Open LNR Combine from Thurso, 502 miles, was sent the same way; there is not much that will beat a semi-Widowhood bird on good days.

Young Bird racing

My young birds are raced semi-Widowhood with wonderful results;

it is not only clean water, clean food, plus good birds that make a successful pigeon fancier, it is the tricks you play with them. Two cocks to one hen, or vice versa. You will find you will have great losses at first, but those that work on it, as I have previously stated, will breed the same. My losses were great when I first tried them, but I hardly lose a bird, young or old, now.

When you are buying birds, go to a successful fancier in your club; if he won't sell, go to any successful fancier who is winning today, not those who won years ago, selling grandsons of champions that have been dead years. You have very little chance of reproducing them.

I must repeat what I've often heard: "When you have mastered the art of feeding you will be on the road to success". My birds are hopper fed summer and winter. I never cut their corn in winter, for that is the time they need it most. I give them plenty of exercise in the early spring to get down surplus fat before breeding. I feed on peas only, with a little maize when I can get it. A lot of fanciers don't like maize, but I fed on maize only one year and I had one of my best years.

There are no hard and fast rules to the game. A lot of fanciers don't like sending yearlings driving; I have won some good races this way, but you must know your pigeons. Birds raced on the Natural system will win, but most successful fanciers have their little tricks. There are no certain types that will win.

Personally I like a good eye. I know this is a tricky subject with most fanciers, but there is a breeding eye and a racing eye. Most of the fanciers who don't believe it, don't know what they are looking for. I am not a great believer in the circle of correlation, but I am in the colour of the iris. My best producers have the violet eye, and breed the red pearl eye, or light violet eye. I think pigeons with a good eye break from the batch quicker when they are nearing home.

I like to train my youngsters with plenty of single-ups, from about 15 miles, picking a certain landmark, making that their main training point. I don't believe in stopping old birds up to 300 miles. I like to keep them at it. If they can't stand the stick up till then, they will never fly 500 miles and win. I am not a wing faddist; I like my birds to moult freely, then they are in good health. I remember one of my good cocks 1729, the year he won money from Fraserburgh, 428 miles, moulted two flights at once in both wings. When I saw that I fancied him all the more as they were half grown.

I like youngsters for the long races completely to finish the body

PLUM, NURP40WGS158.
The father of the loft, winner of seventeen 1st prizes, six 1sts Fed. Sire of seven different Fed winners. Grandsire of many more.

moult, that is to have moulted new bars. I never worry about the wing or tail as I have won good races with youngsters with only six tail feathers and a gap in both wings, but they have finished the body moult. It is obvious that with the covering feathers missing the wing has not got a good downward movement, and the air can get through.

Near to Nature

My birds have an open loft during the racing season. When I let my birds out in the morning, they only fly for about ten minutes, and go straight down in the garden. I often come home and find only a few of them there; the others have been out running.

I like to keep them as near as nature as I can. I use about two cwts of acorns every year and my birds love them. I don't know if there is any goodness in them, but the birds like them and that is the main thing. It also helps the rations out.

I never go in the loft without some sort of titbit, rice, Quaker Oats, baked crusts, anything they like. The main thing is to keep them happy. I never shut pairs up in a nest box to pair them up. I let them out all the morning together, then get them in and let all the other birds out except this pair. The cock will take the hen to his nest box. I let my race birds pair how they like as I believe in love matches, for they race better. Another one of my best cocks 1733 was paired to a tippler; it was his fancy and he won seven 1sts. After a couple of years I paired him to another hen to breed from, but he did not race as well.

Three days before the Morpeth race, which is the Gold Cup race, owing to the wrong liberation time, my birds were all out. When they all clapped off, 1733 was dropping in, he didn't stop but went with the others. I dropped him when they came round and he was 1st Club, 3rd LNR Fed, missed the Gold Cup by three yards. There was a love match for you. This was only one instance. As I have already stated, you must know your pigeons.

An Open Loft

If you are like me keeping only a small number of birds it is very easy to get them much too closely related by breeding heavily from one or two pairs. This close relationship would not matter perhaps if every bird was a champion in its own right, but that, in my experience, it not what happens. The very best put together do not produce a string of certain winners. In fact, sometimes they produce some downright bad pigeons, as I think all honest breeders will confess that.

Reverting to my procedure of about a fortnight before mating; after six or eight weeks on a sparse diet, I then start feeding as much as they will eat, and during this time I try out each pair (in a separate compartment). This enables me to have a final check-up to see if I think I have chosen the mates correctly and also to make sure that they will 'go' together. This, of course, really applies only to the yearlings and new matings, though I let the older pairs have a practice run, too. But it has the additional advantage that when the nest boxes are opened up each bird knows its mate if not its box. I always find one or two pairs are a bit slow in getting the hang of things as far as nest boxes are concerned. Once I have mated-up, as I have said, the loft is permanently open.

When the young birds are weaned I put them into a separate compartment, also permanently open and feed them twice a day. At first, of course, a few old cocks hop in and feed them. Both the youngsters and the old birds seem to enjoy this, but they soon tire of it and I do not think any harm is done, since after filling two or three young birds the old cock goes back to the hopper and fills himself up again.

It does not take long, anyway, for some of the brighter young birds to discover that food is always available in the old bird section. Usually the old birds chase them away, but the persistent ones can always snatch a quick snack. It is here that the small seed mixture comes in. Since the young birds have an open loft they can vanish for very considerable periods, either ranging or just sitting in some field. I always feed them the titbits after my lunch, and they soon get the hang of always turning up at this time, probably following the old birds' example, so that I retain some measure of control over them. I have not been able to discover where either the young birds or the old birds vanish to when fielding. I can only think that they go to many different places, sometimes I suspect, miles away. On occasions, in very dry weather, they return with wet mud up to their 'knees' which may lead a Sherlock Holmes to deduce that they have been on the bank of a river or pond. Sometimes the evidence of my nose points strongly to a farmyard!

Having an open loft has many advantages; the birds can be out early in the morning before I am awake, they can eat any mineral or oddment they fancy, like a wood-louse on the hoof. They can amuse themselves finding twigs for their nests, most of which never get to their boxes, since they forget what they picked it up for and drop it in the loft or on the landing board. But it has its disadvantages. So far I have not suffered from poisoned grain or fertiliser as there is not a great deal of arable land close to where I live. The biggest snag is that one does not know how much flying time per day the birds put in, and since to win in present-day racing, intensive training is necessary, it is difficult to gauge just how many training tosses are needed.

Pairing Up

I mate up about the end of February and breed one round only, which gives me as many young birds as I require and am prepared to look after. Additionally, as Dr Barker advised, I can put aside all thoughts of breeding and rearing youngsters once training and racing is started. This does not, however, preclude me from letting an occasional egg hatch so that the parents can race to a youngster.

I keep very complete records but usually remember all I want to and I find that after mating my hens will lay at about eight days and when they rear a pair of young birds they will lay again at 37 days. When their eggs are replaced with pot eggs (that is, not rearing any young birds) they will usually desert after sitting 22 to 25 days and lay against at 31 days. This I find is a very regular cycle with all my pairs. Almost the only variations are (a) how long a pair will sit before deserting, this desertion may be influenced by the birds having been away at a race, but even if they sit long over time the hens will still lay without maybe any driving, at about 31 days; (b) towards the end of the season, perhaps because of the warmer weather or because the birds are fitter, they may lay a little quicker, in July maybe after only 28 or 29 days; (c) some, but by no means all, older hens mated to an old cocks whose enthusiasm has waned a bit, may be less free-laying. I did not find this frequency of laying influenced much by the birds being away at races, or even by them having a difficult tiring race.

I think most long-distance fliers would consider the end of February too early to mate-up, but the Lerwick NRCC race, which is held as near as possible to the longest day (June 21) is not late in the season.

Apart from this, I should have the greatest difficulty in preventing my hens from laying if I postponed mating until later. Many experts have insisted to me that it is easy to prevent hens laying, but that is not my experience. I suppose I feed too much!

For my later years I did not race young birds seriously, as I am eccentric enough to think that there are other things to do besides race pigeons. I train them with about half a dozen tosses to just over 50 miles including a pretty genuine single-up, which demands a lot of time and infinite patience. If this number of training tosses is thought to be very low, it should be remembered that these young birds have been ranging freely for more than three months. After this I send the whole lot to a race of about 120 miles. This treatment works reasonably well with not very heavy losses.

Winning Lerwick (621 miles)

I figure that a bird that is going to fly Lerwick has to be born with some natural ability and this fairly rapid jumping gets rid of those that have not got a good ready-made 'navigation system'. I do not believe that whatever a young bird does is any indication of how it will perform later on. But a friend of mine, whose opinion I respect, and who has a fine record for long-distance flying, tells me that all his best birds were outstanding as youngsters. This, however, is contrary to my own experience and my observations in following the later careers of what were successful young birds.

My yearlings get much the same treatment as the young birds, flying three or four races up to about 250 miles. This happens to suit my laziness and the Combine programme. I know full well that many yearlings find 350 miles well within their capacity and, in fact, they not infrequently score from 500 miles. I once sent one to Lerwick and got it back, but I do not count that as a success. As two-year-olds I send them to the 450-mile race and then as three-year-olds they should be ready for a crack at Lerwick.

The three hens that I timed-in from Lerwick in 1964 each flew previously two races, one of about 120 miles and one of about 160. I do not think a bit further would hurt, providing the fly was not too hard, but these were the races available on suitable dates, and I prefer to send birds to a convoyed race rather than a long distance to a private liberation. In the seven days preceding basketing for Lerwick, each of these birds flew four training tosses of just over 50 miles each. These tosses can be arranged so that the total time the birds are away from the loft is only about four hours and not more than five. Far from 'taking the stuffing' out of the birds, if everything goes well, I find that mine can actually fly weight on.

If I might make a simple analogy — an Olympic marathon runner does not find himself exhausted by walking down to the local pillar-box and back! Since, to a decent fit pigeon, under weather conditions that are not unfavourable, 300 miles is not even tiring, 50 miles every other day is certainly not a formidable task.

After the race

When a pigeon returns tired from a long race I let it eat exactly what it wants. From my observations a tired bird does not gorge itself, and in any case, a pigeon's digestive system is not comparable to that of a human being and it has time to make some recovery before its crop starts to empty through the digestive tract. If the bird's mate is at home they sometimes enjoy a quick bit of love making, but after that it is probably as well to remove the mate lest it prove too demanding. To show that this system is not harmful I can instance a cock of mine treated in this way that flew Thurso (511 miles) once and Lerwick (621 miles) five times.

I like to get the birds parted by the end of August, and one year I separated each pair as they came up to 21 to 22 days' sitting (and this, of course, is where I started this chapter). To me they always seem by this time to be rather relieved at being freed from the tiresome chore of sitting on eggs every day. I do not breed late breds. After being assured by many experts how good these were, I tried every kind of treatment on them, with only one result in every case — dismal failure.

I have endeavoured to set down my method of Lerwick management, but, of course, it is only possible to give the broad outline and there are certainly many season to season, and day to day variations to cope with changing weather, and to endeavour to snatch success from any favourable opportunity that unexpectedly offers itself. My method really aims at the complete happiness of the birds and my own amusement.

The difference between success and failure in pigeon racing can be very small. But it is certain that in the successful fancier's loft no incident is too trivial to be deemed worth noting and receiving immediate attention if necessary. It is true that "the man makes the

BLUE HEN TOEY, NU61L3213.
Winner of nine 1st prizes, twice 1st LNR Fed. Dam of 11 different 1st prize winners. Sire Champion Mick. Dam Silver Queen.

bird".

Winning Fraserburgh (428 miles) again and again

My good red cock 'Red Admiral' was prepared for the Fraserburgh race which he won 6th Open LNR Combine, 4,812 birds. That was his lot for the year as he was only a two-year-old. He had been on the wing 13½ hours. The next morning after timing him in when I went up to the loft he looked so well, even better than before I had sent him. I made up my mind to send him to Thurso 502 miles a fortnight later, I could not see the four I had kept back beating him. By now the eggs he was sitting on were due to hatch in about five days' time. This meant that he would have a five-day-old youngster for Thurso and this I don't like so the next day I put a two-day-old youngster under him, taking the eggs away and putting them under something the same age. Then two days later I put a youngster a week old under them.

Five days before the race I took the hen away from him where he could not see her; the next day I put the youngster back where it had come from till about seven o'clock at night, then returned it to 'Red Admiral', he wanted to feed it but it had had enough from its original parents, and so I repeated this until Wednesday, the day of basketing. Then I put the youngster back in the nest box at 4 o'clock and I also brought round the hen and put her back in the loft. He was like a dog with two tails. On the Saturday the race was very hard, 15½ hours on the wing, 1st Club, 2nd Open LNR Combine 3,538 birds winning Osman Memorial Cup for meritorious performance. Again I was very pleased to think my trouble had not been in vain, also as he was pooled £2 Open, £2 Section, so I was paid for my work. That was in 1952.

In 1953 I put up one of the best performances, if not the best ever put up in the LNR Combine from Fraserburgh, 5,407 birds, only 18 home on the day from which I had three; 2nd Open, 6th Open, 17th Open, winning over £500 and timed the first bird in London the next morning out of six entries. Again I will tell you how this was done. the first one I timed in was 'Sure Return' he was a winner of eleven 1st prizes. The second one I timed in was my favourite, the 'Combine Cock'. He was a magnificent pigeon, perfectly balanced, a typical Stassart, and always looked well. He had won ten 1st prizes, twice 1st LNR Fed. They were paired to the same hen which created jealousy. A week before the Fraserburgh race I took 'Sure Return' out of the loft, the same time I took the hen paired to the 'Combine Cock' away. The next day the 'Combine Cock' had got the hen in its nest box, I must point out I have no nest fronts on my nest boxes and this pair were next to one another. I left them together for two days then I took out the 'Combine Cock' and brought back 'Sure Return'. Naturally she went straight back to him again for two days, then the 'Combine Cock' for the next day and the next day 'Sure Return'. On the day of basketing the 'Combine Cock' in the morning and 'Sure Return' in the afternoon. One hour before basketing I brought both the cocks back; I sat in the loft letting the hen have a quarter of an hour with each pigeon in his nest box. Of course now and again the other cock would fly into the box where they were, I would gently take him out and put him in his own box.

Racing at dusk

On the Saturday it was a bad day, very bad visibility, light head wind, they were up at 5.30am. By 8pm my phone never stopped ringing with fanciers wanting to know if I had got one and most of the fanciers I spoke to didn't think any would get home on the day but my words to the last fancier were at 8.15: "If they are still flying I would get one", as 'Sure Return' dropped in the pitch dark as a young bird when there were only 12 birds on the day. At 8.25 a pigeon was coming straight out of the north and I knew it was my blue cock 'Sure Return' and he did look well.

He was my number 1 pooled through. I came out of the loft where I had two friends outside in the garden and my words to them were I will get the 'Combine Cock' now and at 8.42 same way a blue cock was coming and I said that's him, the 'Combine Cock'. Sure enough it was and again I said I will get 'Ragamuffin', already a winner of nine 1st prizes, 1st LNR Fed. At 9.27 six pigeons came along the railway bank, almost dark, and I spotted him on the nearside and I called and he dropped out like a stone. His hen had been away for over a week and I had put her back two hours before basketing him.

With the same system he had won all his 1st prizes before. This I have always found, that children will race the same way as their parents especially driving cocks and two cocks to one hen. After this race I did not care if I ever won again — again I had the thrill of a lifetime plus the glory and also once again my tricks had come off and this I think meant more to me than the £500 they had won.

On the night of basketing a very good fancier came to me and said the six blues I had sent were the best he had ever seen go through a marking station. I congratulated him on his good judgement when I saw him at the station for the next race. The same three birds were sent to Thurso 502 miles two weeks later the same way, 14th, 29th, 35th Open nine days in the basket owing to bad weather in Scotland. My first bird was 'Ragamuffin', then the 'Combine Cock' followed by 'Sure Return'. My only disappointment was I never saw them come and they were timed in by a friend of mine as I had booked my holidays previously. They were all grandsons of 'Champion Plum'.

Hard Luck

I work my pigeons very hard; I don't like big jumps, I send them to every race if fit, a good pigeon thrives on good work. Most fanciers have birds come home two or three days late from a race and don't send them back for two weeks. If they are fit I send them back the same week and if you are going to win with them this is the time. My good blue chequer cock '1729' came home on the Tuesday following a race. He was a bit down, but on the Friday he was my fittest pigeon. As we had a good Open race that weekend I had several in line for it but I made him my best pool bird. He won the race by four minutes and never had another night out. Most of my best pigeons I have ever owned have made a mistake somewhere along the line. My good blue cock 'Champion Mick' had a night out on a very cold east wind, two weeks later 1st Fed, 4th Open LNR Combine Berwick, 4,668 birds. I used to take his hen away for a week and put her back two hours before basketing and we won seventeen 1st prizes, four times 1st LNR Fed.

My good mealy hen 'WG78', daughter of 'Red Admiral', was my most consistent young bird, winning two 1sts and several other positions. She made mistakes in the young bird Combine and returned in March the next year still with her rubber ring on which was almost perished. She had definitely not been in. I gave her a couple of races that year and stopped her. The next year after having a perfect moult she was my best pigeon for three years winning three Berwicks 300 miles, three Fraserburghs 428 miles and three Thursos 502 miles. Each time a good position in the LNR Combine. When she was four years old I intended taking her off the road but that year I had trouble with my young birds after coming home from my holidays. My son-in-law must have let in a stray youngster that must have had roup so I culled most of them and stopped racing.

That made me short of a racing team next year so I thought I would give the mealy hen one more year on the road with a good gamble from Thurso in mind. I gave her about five tosses from 35 miles and jumped her straight into Berwick 300 miles. She beat my others by five minutes winning 1st Sect, 7th Open Combine 4,750 birds. That year we flew Thurso before Fraserburgh this time my money was on her she was 1st Club, 24th Open Combine and all pools. A fortnight later 1st Club, 33rd Open Combine Fraserburgh. That year she almost won the Thomas Long Trophy on her own. Her prize and pool money totalled over £600.

My good blue chequer cock 12685 won four young bird races, three times 2nd Fed, once beaten by his loftmate 1st Fed. As a yearling he won four 1st prizes in a row, I sent him to an Open race of 200 miles and the race turned out to be a bad one, the weather deteriorated after liberation and only a few got home on the day. I have myself to blame as I was race adviser! He homed the following spring still with the rubber on but he had found somewhere where there was plenty of food as he had had a good moult. Again I am sure he had not been in.

So I raced him right through to Thurso winning 1st Northallerton, 1st Fraserburgh, 1st Selby, 1st Thurso only three birds home in the club and the £2 Section, £2 Open pools being nearly 14 hours on the wing and when he dropped after flying round for three minutes he looked a picture. I could go on with other good pigeons that have made that one mistake and have made champions. The point I am trying to stress is never condemn a bird that makes one mistake. If they have had a bad moult whilst away let them lay by for one year and I am sure they will repay you the next year.

Fresh air and a problem

I believe in plenty of fresh air, my floorboards were put down half-inch apart. In the summer they open up to nearly one inch. This cost me the Gold Cup, I dropped a rubber ring and it went through the floorboards and under the loft, so under the loft I had to scramble and when I got under I could not see it, so back in the loft to see where it was. There it was between two bricks and I was beaten by a decimal point to the runner-up. Both the front and side of my loft is open and when visited by the Belgians they say "Too cold. Put shutters up". I agree with them, most of the lofts in Belgium are closed in, just enough ventilation where the birds enter from a race. But I like fresh air and I have shoveled heaps of snow from my loft. One year it had blown all around my nest bowls and I still bred my best young birds.

My racing loft has 30 nest boxes, five high six wide, no box perches every bird has his own box. This stops fighting and gives enough room so they can have a good moult. I have two pieces of half-inch gas barrel tubing from the roof to the floor so they cannot jump from one box to the other. The boxes are 14 inches square with no nest fronts, the loft is 14 feet long, five feet wide, seven foot at the front, six foot at the back in two compartments with an aviary at each end and a four foot verandar the length of the loft on the front. One of the biggest mistakes in the pigeon game is most lofts are too wide, you cannot control your birds in a loft they can fly round in. I like my birds all round me where I can pick them up whenever I want.

It makes me laugh when I see a fancier wave one hand and grab his birds with the other when they are on the perches and nine times out of ten miss them. They look at me and call him or her all the names under the sun when it's not the pigeon's fault but bad loft arrangement and no stock sense. I always move about quietly when I am in the loft and am always giving them titbits and talk to them. I get most of my hens to swoop up to me, these are always by best stock birds and racers. I have had two hens paired to the same cock and they have won 1st & 2nd Fed but you need plenty of patience.

Guarding the nest

My cocks guard their nest box, when I go to clean out the boxes every morning I have a job to get the scraper inside. They are always ready for a fight, these again are my best racers. My good white pied cock 'Omo' when he went in his box from a race I had a job to take off his rubber ring for him pecking me — no pigeon was flown harder twice a week from races up to 300 and 400 miles, for three weeks and he was unbeaten. Good pigeons as previously stated thrive on hard work and on race day I like to see every pigeon come, the late ones get the same titbit as the early ones. I never mind how late they are so long as they come racing and fit. By this method I have found my best long distance birds. The good mealy hen 'WG78' from one of the short races would drop when it was nearly dark at night and from then onwards she was right.

During the young bird racing I take them 15 miles away and an hour before darkness let them go in twos and threes and have got back home to see most of them drop in nearly dark. Those that have a night out are hard to beat at the weekend. In 1949 my daughter and a friend used to take a dozen youngsters on a bus ten miles away and let them go in twos every five minutes just before dark for a week. It has been so dark at the home end that they have missed the loft and dropped in the garden where I have picked them up and put them in the loft. Three of these youngsters in 1953 from Fraserburgh were 'Sure Return', 'Combine Cock' and 'Ragamuffin' but I think this type of training today would be suicidal with the amount of television aeriels there are.

One hears so much about clashing year after year, this has never worried me, good pigeons won't clash and go as they seem to think the other way. If they do I am sure this sorts the wheat from the chaff. The trouble with the sport today is there are too many birds at the race point, too many young birds liberated together, 6,000 and sometimes 8,000. Most of them have hardly been trained. "They'll be alright the others will bring them home." I have heard this said so often in my own club. I know this is a sport for everyone but the Classic races entry fees are too cheap. They send birds that have been away two or three weeks just because they have a Combine ring on. If the race fees were more they would think twice about sending them, and you would get the best from each loft. I find I am wandering away from the title of my chapter into pigeon politics and am sure you read and hear enough of this at your meetings.

Chapter XV
Keeping things the same

I have an old jumper and always wear a cap in the garden even when it is very hot I put on the old jumper and cap to time them in. Some years ago I forgot to do this and I had a good pigeon but when I went in the loft to pick him up he flew out like a rocket. I came out of the loft and he was flying, my wife had seen what had happened and she said it's because you have not got your jumper and cap on. I went and put them on and he dropped and went straight in and I had no trouble to pick him up — he was just beaten. I can assure you pigeons are not colourblind as some fanciers think. They don't like bright colours or something they have not seen before. You hear the noise they make you would think there was a cat in the loft, even when you walk up the garden wearing something strange.

Why fanciers put a light in their loft I don't know as I never go in my loft after dark unless for some unforeseen circumstance. Pigeons need all the rest they can get during the racing season so make sure they are not troubled by cats peering in the loft from some outer building. If you are in a position to do what I did, put two foot interwoven fencing on top of your fence both sides then two feet of wire on top of this. You may say I have a dog that keeps the cats away, my old birds have an open loft as soon as I pair them up and are always pecking about in the garden. If you do make sure they don't go down the part of the garden the birds go down.

I remember having a flat top corn bin in my old bird racing loft where I kept my corn and every time I came home from work there was a heap of twigs and straw on top of it. So I threw the twigs outside the loft but the pair were most persistent and as they were two nice yearlings which I had hoped to do well with. I took the corn out of the bin and let them have it. They were my two best pigeons that year but I can assure you they did look peculiar sitting there. Another pair of yearlings after rearing their first nest in the loft wanted to nest over my work-bench in my work-shop as this was quite near my loft. The cock used to bring the hen to the door, he would fly up on the bench and on the shelf and start to call the hen but she was a little shy, until one day I went out to do some work and when I came home they were both on the shelf on the bench. As they had already reared the one and only pair of youngsters I wanted before racing I gave them a nest bowl. This was a week or so before racing and they used to fly in and out bringing nesting material. The hen laid and soon forgot I was there. I used to trap them inside the loft and no sooner had I taken off the rubber ring than they were in the workshop to their eggs. They won three 1st prizes, the cock was also 4th Open LNR Combine, 4,079 birds and well pooled.

Strange nesting places

Some years ago I had a young cock and hen that used to come back late in the evening every day and I wondered where they used to get to all that time till one day after letting them out I saw two come back from the others that had gone running and drop straight on the house and then drop on my next door neighbour's attic windowsill and disappear. I went and saw that the window was open three or four inches at the bottom and I knew it was never used as they often told me it was their junk room, until a few days later she called me and told me they had given her a scare when she went up to get something.

She said she would shut it down. As this was only a few days before the Young Bird Combine I asked her to leave it open and this she agreed. She said there was nothing up there they could hurt and I put my money on them. The cock was my first pigeon 12th Open LNR Combine followed by the hen a minute later 27th Open. A few minutes after I had timed in all I required they had gone back to the junk room. The cock was verified the next morning as he was the earliest at the station, I had to keep him in or else I would have had to go next door to get him. I am only trying to prove the peculiar places some pigeons want to nest.

Another good thing for pigeons is sprouting corn this I do before the Classic races. I wet two sacks and lay one on the lawn in the garden and put a tinful of peas on top and cover them with the other one, in three days they will start to sprout and then I throw them on the lawn for them to pick at. I make a habit of boiling my drinkers each weekend in an old gas copper which I have in the conservatory. I put a bucket of water and the drinkers in and bring to the boil and scrub not only the inside but the outside. This is most essential, then I scrub my bath out with the water from the boiler. This may seem a lot of hard work and it is but a man only gets out of his birds what he put in.

Make a note of birds that are good nesters, meaning those that are all the time picking up straw from the garden which you have put down, and build a good nest. Not only are they your best racers but they will breed your best youngsters. I like runner bean stalks I leave them to dry out and pull them down when they are going to nest and break them up into about nine inch pieces. They like these better than straw. I sometimes go to my brother-in-law's allotments and collect his or anyone else's that will let me have them. When collected and taken in the nesting boxes it helps the air to get round the eggs and helps to breed good youngsters. Watch the hens after laying, those that still keep bringing in the straw, those are also my best hens likewise cocks as they are the keen ones.

Some of my birds build their nests nearly up to the roof of the nest box. This happened to me as I wanted the hen sitting about eight days and as I knew she was due to lay I could not understand why she had not done so, and the cock had stopped driving her. After a week or so I went up to the loft just as she was laying and I saw the egg drop over the side of the nest. When I moved the bowl away I saw the other two eggs I was waiting for her to lay so now I always look behind the nest bowl when I have a hen overdue to lay. Keep plenty of nesting material down this helps to keep them happy. This is one of the main things with pigeons is to give them something to do. A handful of linseed after they have reared their youngsters thrown over the lawn, this they will peck about finding it for hours. It is surprising what they find to do when you have an open loft and the things they use for nesting; screws, six inch nails, pieces of chain and buttons and also your car keys which I lost and found in a nesting bowl after I had weaned a youngster.

Ashes and Water

I bred a youngster for a friend of mine and a few days after he had taken it away he said it was a bit off colour so I told him to cull it and I would breed him another one. He said he would give it another week or so and see how it went. A few days later he rang me and said as he picked it up something pricked his finger and when he blew away the feathers from that part he felt something sticking out and he told his wife to get a pair of tweezers and she pulled an inch and a half sewing needle from his crop. The pigeon flew very well. It was a needle which my children had dropped while sewing out in the garden and this pair had taken it into their nesting bowl. When I burn all the twigs and pieces of wood in the garden I throw the ashes over the lawn and they peck for hours among it, picking out the charcoal. Make sure the wood has not got tar or creosote on it or that what has been painted with emulsion as this is harmful to them. This is the only thing I put on my lawn. I do not use fertilizer in the garden where birds go pecking about this also does them no good.

Put the bath out two or three times a week when rearing is a good idea this helps the incubation of the eggs and stops youngsters dying in the shell. I never let them bath a week or so before the long races as they need all the bloom and powder on the body to get through a belt of rain. Thus helping them to do more miles before getting too wet and having to drop. You can see the bloom which comes off them, every time you bath, on the top of the water. This is Nature's way of stopping them from getting too wet. I never leave my birds out when it is damp and cold, I do not mind leaving them out in the rain if it is warm and when it is raining hard I put the bath out empty to catch the rain water for them to drink.

Semi-Widowhood

There have been several books written on the Widowhood system and from those that I have glanced at are too complicated with the double nest boxes with a wire slide between and do not let the cock tread the hen. I have flown this type of system for years and have no double nest boxes, I have not even got nest fronts where you can put the hen in and let the cock sit on the outside box. I have found the semi-Widowhood system equally as good. It does not need commonsense to know the amount of winning some of my good cocks have done that this is not only on eggs so I will try to explain and try not to make it too complicated.

First I let them rear one round of youngsters then fly them on their next round of eggs. Try and get four races out of them in this condition. Make sure they are pot eggs meaning boiled eggs as previously mentioned. Some pairs will sit well overdue, this is what you want. If you find they are on the verge of deserting them after getting only three races out of the cock give them a youngster so that you can get another race thus making the four required. This lets them find a line and gets them used to what it is all about again and to have now steadied down.

The Sunday after the fourth race, that is presuming the race was flown on the Saturday, take the hen away and put her where he cannot

OMO
Winner of ten 1sts including three Open races. Flown Thurso (502 miles) four times, Fraserburgh (428 miles) four times. Son of Red Admiral and Scottish Lass.

see her. This is of most importance, after two or three days he will get over being parted and should be one of your most active pigeons outside the loft. You must have no spare hens in the loft which he can pair to. So you should have put them in the same place where you have put his hen. Also take away his nest bowl from inside the nest box leaving the box empty. All the time watch his hen does not pair to another hen as this sometimes happens so make sure there are no nest bowls lying about. The following week's race send him spare meaning without his hen. I have even won with them in this condition but he is not ready yet. I must point out I am only speaking of one cock but I work three or four at a time.

On the Thursday or Friday the day of basketing for the next race I send all the cocks that are on the system, in the morning to a 35-mile toss, of course weather permitting. Then about 4 o'clock I put in the nest box a clean nest bowl with sawdust and sand also a handful of straw on top. If you have an open loft make sure the cock sees what you have done. If they are out, call them in so they can see the nest bowl in their boxes. They will go straight in their boxes and become very excited. About an hour and a half before basketing him for the race bring round the hen. Make sure the cock is out of the box so he does not see you put her there. In my case I drive the cocks outside the loft and put the hens back in their boxes and get the cocks straight in then drive them in the boxes where the hens are waiting for them. He may just sit on the nest bowl calling her, he may tread her but I have found this makes no difference, in fact I am sure this may make him keener.

When you have basketed them up for the race put the hens back where they have come from and on the day of the race have them in the basket ready to put in the nest box directly they arrive home. This is of the utmost importance and leave them together till the following morning, when you take the hen away again where he cannot see her and repeat the same thing for the following two weeks. On the third week leave the hen with him and let her lay and sit for 21 days on pot eggs then take her away again. If you get one that works on it wait till there is something to be won and repeat the system. I would point out the hens you use must not be your best but those you do not intend racing. They must be kept only for the system and at the same time when you have flown the cocks three times on it you must have some more cocks ready to take their place by previously having other hens away.

Semi-Widowhood Cocks

I must mention that the cocks you have in line for this must have flown consistently and must be paired up early in February or so they have reared a pair of youngsters and are on their next eggs for the first race. If as previously mentioned you do get one which works well on it don't be too greedy as three times in one year on it is enough. Do not worry about the other pairs you have sitting in the loft. The hens from them will not be interested in your semi-Widow cocks if truly mated or are a love match. You may think why the cocks must not see you put the hen back in their nest box. This is where you make him think he has not been alert enough to have noticed her and sets his brain thinking of her more actively. That is why I like to give them a 35-mile toss on the day of basketing as he may think she was in the box when he came home and he did not see her there as nine times out of ten they do not go straight in the loft as there is nothing to go in for but the next time he goes away he will be more alert.

The same thing applies to the hen mating up to another hen she will not be so keen to see him when they first see one another again and she will be more inclined to go back to the other hen. It is the small things that make this system work and I have won half of my 1st prizes this way. With the 35-mile toss you give them you do not have to be there to see them in. It does not necessarily have to be that distance, ten or even 15 miles will do.

The same system I work with young cocks but wait till they have cleanly body moulted before I get them paired to an old hen for about four or five days. Then I take the old hen away and put her where he cannot see her for about three or four days and bring her back half an hour before basketing leaving out the 35-mile toss on the day of basketing. I have three nest boxes in my young bird loft and the young cocks who commandeer these are the ones I put on the system. If I have a young cock that is flying well but is only on a box perch I put an extra piece of wood on the front to take a nest bowl. This I do by getting a piece of wood the same width as the perch and about five inches wide and screw it on with brackets underneath the other one, and thus making a box perch wide enough to take a nest bowl where he can take an old hen up to, but make sure you fill the bowl up with dry sand only so it won't tip over or wobble or be unsteady. I can assure sand only so it won't tip over or wobble or be unsteady. I can assure you these small things are most essential. In 1969 I did this with a young cock but he did not take an old hen up to his perch, but a young hen; this was a week before the LNR Two Bird Championship Club, they were 1st & 2nd winning £108. So once again I was paid for my trouble for putting that extra piece of work, and again one of my tricks had worked and this alone gives me great satisfaction with my pigeons.

The Jealousy System

The Jealousy system I have mentioned previously also applies with two hens and one cock, but I do not emphasise this too greatly as one of the hens will have to be kept in for a couple of days at a time where the cock cannot see her thus not getting the exercise required to keep her fit. But I have had some good results with them, not only hens but the cocks as well. An hour before basketing I shut the cock outside the loft on his own with the two hens for half an hour keeping the other birds in the loft then I get them in and stand in the loft to stop the hens fighting over him. This is most essential as they can do damage to one another. This is for the remaining half an hour. I stop them fighting by putting some wire in front of the nest box after I have put one of the hens with the cock. I have some perches in front of the nest box where the other one will stand outside and try to get in. After about five minutes I take the hen out of the box and replace with the other hen. At the same time replace the wire front and repeat several times. Then just before you are basketing them for the race let them go in the box together. Directly they start to fight part them and basket up for the race likewise the cock. The wire fronts I mentioned were made to fit every box not only for this purpose but to put in front of the nest boxes whilst a pair is away sitting, as the others will pull their nest about and disturb it. I like them to come home and find it the same. Sometimes I warm their eggs before their arrival if I want them for the next week's race. If you disturb a wild bird's nest they will desert it, the same applies to racing pigeons, and you won't find the best conditions your birds fly to. Also it will mean your hens will have to lay again and it is not good for them to lay again too quickly. As I have said it's the little things that make the successful fancier. Directly you have parted the two hens make sure they do not go in the same basket likewise when

MARGUERITE
Winner of eight 1sts, 8th Open London NR Combine Fraserburgh, 4th Open London NR Fed Lerwick.

you take them to the club. I use a separate compartment basket and take each hen out directly after one another so they go in different club baskets for the race. This also applies to your jealous cocks, again it is the small things that are most important to get the best results. I do this only once with hens and have had the two hens together from the race followed by the cock to take 1st, 2nd & 3rd Fed with over 8,000 birds.

Racing and Showing

I have had the honour to judge at most of the big shows and always ask for a racing class as I do not like going through a class of show pigeons they have got nothing for me compared to what the racing pigeon has. Their heads are too round and have not got the intelligent look or the impression they would fly their hearts out to get home. They give me the impression they would drop on the first house when the going got tough. Nevertheless there are a lot of fanciers that get pleasure from them and I would never turn one down in a class if it was put down right and believe me it would have to be without any kind of frets and not a spot of dirt anywhere. The first thing I look for is cleanliness no loft muck on the feet or tail. The tail should look like one feather on top of one another and plumage should stand out like the pigeon is made of marble. When judging I give the marks to the pigeon I would like to take away myself provided it is put down as I have mentioned.

I am not a colour fadist. Some judges like reds or mealys, some like blues or blue chequers some judges do not like grizzles. I remember giving a grizzle hen a 1st prize in a local show and I heard the whispers going around, "Baker's given a grizzle a 1st". The next year this hen swept the board in that club being several times well up in the Combine. I have never made showing in the winter months a habit, I have sometimes put a few out in my own local club more or less for support but I do not like taking birds out of the loft at night especially my best racers. I have seen the way some judges handle them, I remember seeing a judge getting out a pigeon tail first, it is a good job they were not mine or I would have told him a thing or two like the time when I had worked on my pigeons for a Combine race and had put quite a lot of money on six pigeons. I gave my pigeons to the marker who got the first one out tail first and I told him to get them out the right way, head first. The next marking night he got my pigeons again to race ring and when he saw the name he said: "Baker's pigeons. Get them out head first", so I said I am glad you are learning.

I remember when I was first asked to mark pigeons at a Combine station, how proud I was, today you need a horse whip or they want paying to get their help. You get the same fanciers each year that do the donkey work, still I suppose they are the men you want. They are the pigeon men and they do it because they love pigeons and they do it right. The others even drop their baskets of pigeons and make straight for the bar and when the clocks are set they are away, leaving the same men to carry the birds out to the lorry. I soon stop this; I wait till all the baskets are on the lorry and then we give them a strike! It is a pity I cannot do this for the old bird races but they are away before the clocks are set.

I again find I am running into pigeon politics and have thus run out of steam regarding my ways and means of my life long study of the racing pigeon. As I have said if I live my life over again I would still be learning as the racing pigeon is something we can never stop learning about and each year I find I have learnt more about them and no one knows it all. And if there is an after-life my wife says when I have the down feathers on my hair "you will come back as a pigeon". If I do I hope I get in a loft where someone loves pigeons like I do and always remember we can all be good winners but also be a good loser and I hope when you have read this book you have something to help you be as proud of your pigeons as I am mine — good racing with the gamest birds in the world.